Broadband Growth
and Policies
in OECD Countries

OECD

ORGANISATION FOR ECONOMIC CO-OPERATION AND DEVELOPMENT

The OECD is a unique forum where the governments of 30 democracies work together to address the economic, social and environmental challenges of globalisation. The OECD is also at the forefront of efforts to understand and to help governments respond to new developments and concerns, such as corporate governance, the information economy and the challenges of an ageing population. The Organisation provides a setting where governments can compare policy experiences, seek answers to common problems, identify good practice and work to co-ordinate domestic and international policies.

The OECD member countries are: Australia, Austria, Belgium, Canada, the Czech Republic, Denmark, Finland, France, Germany, Greece, Hungary, Iceland, Ireland, Italy, Japan, Korea, Luxembourg, Mexico, the Netherlands, New Zealand, Norway, Poland, Portugal, the Slovak Republic, Spain, Sweden, Switzerland, Turkey, the United Kingdom and the United States. The Commission of the European Communities takes part in the work of the OECD.

OECD Publishing disseminates widely the results of the Organisation's statistics gathering and research on economic, social and environmental issues, as well as the conventions, guidelines and standards agreed by its members.

This work is published on the responsibility of the Secretary-General of the OECD. The opinions expressed and arguments employed herein do not necessarily reflect the official views of the Organisation or of the governments of its member countries.

Corrigenda to OECD publications may be found on line at: *www.oecd.org/publishing/corrigenda*.

© OECD 2008

Foreword

In February 2004, the OECD Council adopted the *Recommendation of the Council on Broadband Development* (reproduced in Annex A). The Recommendation calls on Member countries to implement a set of policy principles to assist the expansion of broadband markets, promote efficient and innovative supply arrangements, and encourage effective use of broadband services. The Council instructed the OECD Committee for Information, Computer and Communications Policy to monitor the development of broadband in the context of this Recommendation within three years of its adoption and regularly thereafter.

This report has been prepared in response to the Council mandate. It examines broadband developments in OECD countries since the adoption of the Council Recommendation and highlights policy challenges that remain. It also outlines emerging broadband issues that may need future policy attention and which may eventually require a revision of the Recommendation. The Secretary-General invited the OECD Council to adopt the draft conclusions of this report and in April 2008 the Council agreed to the report's declassification. The report and its conclusions have also provided background information that has assisted preparation for the June 2008 OECD Ministerial meeting in Seoul on *The Future of the Internet Economy*.

This report was extensively reviewed by and received inputs from the OECD Committee for Information, Computer and Communications Policy and national delegations. Contributions were also made by the OECD Directorate for Education (Centre for Educational Research and Innovation), the Directorate for Public Governance and Territorial Development (E-government Project) and other members of the OECD Secretariat. The report was prepared by Taylor Reynolds and Sacha Wunsch-Vincent under the overall direction of Andrew Wyckoff, Graham Vickery and Dimitri Ypsilanti, all of the Directorate for Science, Technology and Industry.

Further reports on broadband and the digital economy can be found at www.oecd.org/sti/ict.

Table of Contents

Main Findings and Policy Suggestions: Monitoring the Recommendation of the OECD Council on Broadband Development

Broadband not only plays a critical role in the workings of the economy, it connects consumers, businesses, governments and facilitates social interaction. The Recommendation of the OECD Council on Broadband Development recognises this growing importance of broadband and its principles have been instrumental in fostering broadband development.

Over the previous three years, policy makers have followed the Council's Recommendation and implemented many of the suggested policies. Broadband policies are now a vital part of broader ICT policy strategies and are now receiving the same attention as other key economic policies. The principles should also prove useful for non-member economies.

Yet, the monitoring exercise also reveals that there is still scope for OECD countries to improve broadband development. Some principles of the Recommendation need renewed attention and some OECD countries have fared better in their implementation of these principles than others. A number of new issues have been identified which need to be added to the existing principles in a future review of the Recommendation.

Positive market and policy developments

The development and use of broadband has flourished in most countries since the Recommendation. Since December 2004, broadband subscribers in the OECD have increased by 187%, reaching 221 million in June 2007. Broadband is available to the majority of inhabitants even within the largest OECD countries. A number of countries have reached 100% coverage with at least one wired broadband technology and up to 60% with coverage by two. Wireless Internet connections at broadband speeds are also increasingly available and are particularly important in underserved areas.

As broadband connections proliferate, connections are faster – and less expensive – than they were just three years ago. The average speed of advertised connections increased from 2 Mbit/s in 2004 to almost 9 Mbit/s in 2007. Prices have also fallen. Between 2005 and 2006 the average price for a DSL connection fell by 19% and by 16% for cable Internet connections. Broadband is also affordable in most OECD countries. The price of a broadband subscription in 20 of the 30 OECD countries was less than 2% of monthly GDP per capita in October 2007.

Data on penetration, price, speed and usage of the Internet highlight how member countries have promoted competition, encouraged investment and worked together with the private sector to increase connectivity. Coverage statistics and penetration rate data show that operators and governments have made great strides extending broadband to rural and remote areas. Satellite services are available in even the most remote areas of many OECD countries, although these tend to be more expensive relative to other access technologies. Many governments have also implemented broadband demand aggregation policies to bring connectivity to rural areas. High-speed wireless/ mobile Internet connections are increasingly available as an important option for users. Discussions have begun concerning how best to measure and compare connections across countries.

On the demand-side, OECD countries have focused on increasing the uptake of installed capacity, electronic business, digital delivery and broadband applications. Promoting the general ICT business and policy environment, fostering innovation in ICT (including R&D) as well as ICT diffusion and use (including e-government) have been priorities. Likewise, ICT skills and employment, digital content and promoting trust have been key concerns.

In particular, OECD governments have implemented demand-based approaches for spreading broadband access. Policy makers have made particular efforts connecting schools, libraries and other public institutions. Overall, these policies have led to increased use of broadband across the board.

Since the spread of broadband, traditional Internet activities (*e.g.* obtaining information) have intensified. New kinds of – often increasingly participatory – Internet activity and content-rich broadband applications have also been on the rise. Higher data-intensive applications are on the horizon, *e.g.* streaming high-definition video and TV, new peer-to-peer applications, health or education applications, virtual conferencing, and virtual reality applications. Emerging usage trends such as the migration towards user-created content and social networking will stimulate further opportunities but will also present challenges for policy.

OECD governments have also fostered broadband content and applications, for example, by acting as model users, by promoting e-government services and broadband-related standards, by putting content online and by supporting the development and distribution of digital content by other players.

OECD governments and industry have also put into place regulatory measures to promote a culture of security. On the consumer protection side, OECD countries have focused on developing awareness campaigns to educate consumers about risks to Internet security; they have also instructed consumers on how to protect themselves against fraudulent practices.

Areas which need more attention

There are some key policy areas highlighted in the Recommendation that need more attention.

There are still substantial differences in broadband access and use among the OECD countries. Levels of competition among Internet service providers vary among the different OECD member countries and also between rural and urban areas within each country. Prices for Internet access in some markets remain high and users may have a very limited choice of broadband providers. OECD policy makers can do more to promote efficient competition in some markets. Governments that have chosen to focus on infrastructure-based competition must create a competitive market environment that provides investment incentives for competitive operators and incumbents. Governments that have historically relied on unbundling for competition will need to evaluate the role and future of unbundling in next-generation networks, and should also facilitate infrastructure-based competition.

Furthermore, there exist specific problems with broadband within OECD countries. While the number of broadband connections in rural areas has increased, the qualitative aspects of these connections vary significantly than those in urban areas.

There are also a number of important issues to do with broadband supply in OECD broadband markets which are not covered in the existing Recommendation. Debates over whether Internet service providers should be able to prioritise or limit certain content and data over their networks (commonly referred to as "network neutrality" debates) are spreading across OECD countries and even across platforms (fixed to mobile).

The Recommendation provides little guidance with the exception of promoting competition in markets. Policy makers also face questions about the future of universal service. The Recommendation gives some guidance on the role of governments and the private sector in promoting connectivity. However, questions remain on how or whether universal service will be adapted for high-bandwidth use, particularly given the Recommendation's emphasis on technological neutrality.

Significant differences in the uptake of broadband in businesses, schools and households still exist among the OECD countries; some with far lower use levels than others. Particular attention needs to be paid to the broadband use of small-and medium sized enterprises and particular socio-economic groups.

The monitoring exercise has also shown that the evolution towards broadband applications and use is only now gaining in speed, and that many services are still in their experimentation phase. The goal of "broadband applications anywhere, anytime and on any device" has not yet been achieved, and commercial online broadband content services are only slowly emerging, in particular, in the areas of audio-visual content, although there are exceptions. As consumers are demanding more advanced content, faster upstream bandwidth is becoming essential for further development of the information society. Advanced mobile (wireless) broadband services and associated mobile content have yet to develop in OECD countries whose access is still largely PC-centric. Furthermore, there is still substantial scope for OECD governments to put more content and e-government services online.

Importantly, OECD firms and governments are only just beginning to realise the full potential of broadband when it comes to advanced broadband applications. The use of broadband in education, for tele-work, for e-government services, energy, health (tele-medicine), and transport (intelligent transportation systems) is still in its infancy. Organisational and institutional barriers hamper the necessary innovation and structural changes needed and leave many OECD countries struggling to move beyond pilot projects. The notions of ubiquitous networks, broadband-based home management, and other new forms of broadband use have yet to develop and diffuse.

A number of broadband-related security threats have emerged in OECD markets over the last three years. The transition to fibre connections and symmetric bandwidth will make these threats more virulent. New or more pronounced consumer and privacy issues are transpiring with broadband's "always-on" connections and its participatory features.

The Recommendation has also highlighted privacy enforcement and consumer protection, both of which warrant policy attention.

Devising balanced regulatory frameworks, especially in fields such as intellectual property rights (IPRs) will be a continuing challenge for governments.

Governments will have to invest in R&D that promotes broadband infrastructure, applications and content. The development of broadband research networks and their use can be developed further.

Finally, only a few countries have specific broadband policy assessment and evaluation activities which would allow them to carry out existing broadband plans in a more effective and accountable manner. Internationally comparable broadband metrics are needed to meet this goal.

Policy suggestions for the way forward

This monitoring exercise of the Recommendation has led to the following policy suggestions.

Evolution of broadband

* The regulation of new broadband connections using fibre to the end user will likely be the subject of considerable debate in the next few years. The pressing question is whether fibre optic cables extending to homes, buildings and street curbs should be regulated in the same way as traditional copper telephone lines. As new fibre connections may fall outside existing regulatory frameworks, a re-evaluation of existing policies may be required. Regulators should consider whether network architectures still relying on portions of the historical copper telephone infrastructure should be treated differently from new all-fibre networks.

* Regulators and policy makers are increasingly concerned about fostering competition on next-generation broadband networks. Some are examining the functional separation of the dominant telecommunication provider into two units, one which handles the physical lines and the other which provides retail services over the lines as a way to ensure fair and non-discriminatory access to "last mile" infrastructure. The results of functional separation, particularly on investment, are still far from certain and warrant significant research. Regulators should actively consider other policy options at the same time, which may provide similar outcomes – such as requiring operators to share the internal wiring in buildings.

* Broadband connectivity has improved but significant divides remain between rural and urban areas. Wireless technologies will certainly play a role in connecting some of these areas but there will likely be more

demand for high-capacity fibre to reach as widely as possible into these areas in order to feed wireless connections. Governments need to help ensure that all citizens have access to very-high-speed broadband networks.

• Competition among providers of communication technology has always been a key goal in OECD communication markets so that Internet subscribers in urban areas have a choice between wired providers and wireless options. However, policy makers should reconsider whether promoting this kind of competition is a realistic goal for rural and remote areas, which may only have one high-speed provider.

• Technological neutrality features prominently in the Recommendation but is not yet a reality in OECD markets. Unbundling requirements on fixed-line operators and local cable regulations are examples of the technological bias still pervasive in OECD countries. With the move to next-generation networks, policy makers may need to re-examine whether technological neutrality is still an efficient policy structure.

Government intervention with respect to broadband infrastructure

• The private sector should take the lead in developing well-functioning broadband markets, but there are clearly some circumstances in which government intervention is justified. For example, connecting underserved areas and promoting efficient markets.

• Governments need to actively look for ways to encourage investment in infrastructure. Civil costs (*e.g.* building roads, obtaining rights of way) are among the largest entry and investment barriers facing telecommuni-cation firms. Governments should take steps to improve access to passive infrastructure (conduit, poles, and ducts) and co-ordinate civil works as an effective way to encourage investment. Access to rights-of-way should be fair and non-discriminatory. Governments should also encourage and promote the installation of open-access, passive infra-structure any time they undertake public works.

• Governments could also help co-ordinate map-making of network routes as a way to encourage the rollout of smaller networks in need of inter-connection. Improvement in the overall investment climate in a country should also benefit providers wishing to roll out new networks.

• Governments should not prohibit municipalities or utilities from entering telecommunication markets. However, if there are concerns about market distortion, policy makers could limit municipal participation to only basic elements (*e.g.* the provision of dark fibre networks under open access rules).

* Any government intervention in markets that involves funding should follow a set of basic rules. Requests for proposals should be technologically neutral and simply specify the minimum criteria for the project. Any new infrastructure built using government funds should also be open access – meaning that access to that network is provided on non-discriminatory terms.

* Access to spectrum remains a significant market barrier to wireless broadband provision. Policy makers should adopt more market mechanisms to promote more efficient spectrum use.

Broadband diffusion, use and policy developments and recommendations

* Certain OECD countries have significant scope to renew efforts to promote broadband deployment and use in public institutions, businesses, households and governments.

* Differences in income, education, as well as gender are factors influencing the uptake and use of broadband in OECD countries ('new use divides'). Such factors need to be better understood and addressed. Sustained efforts to improve ICT and media skills and to foster relevant training are also needed.

* OECD governments should continue promoting the business use of broadband and e-commerce. The imposition of national boundaries on the Internet is a barrier to progress and threatens the positive expectations of the Internet as a global trading platform. Innovation in the area of new web-based services and moves towards more advanced business applications should be encouraged. Studies and policies should focus on the remaining bottlenecks and remedies.

* The business- and user-centric innovation spurred by broadband networks in business but also social and cultural areas needs to be sustained. Governments should focus their attention on improving metrics and analysis to better understand new usage trends, their impacts on the economy and society as well as policy.

* There still remain a number of bottlenecks in the deployment of broadband services and content. Most of these will be resolved by the marketplace. However, governments can also help by providing a forum to resolve issues. Activities supporting the development and distribution of digital content, and policies ensuring competition and innovation in broadband services should be intensified – especially as they relate to R&D. Improving framework conditions, skills, common standards, and

facilitating cross-industry collaboration is also necessary. With increased digital convergence of broadband and media services, the regulation of digital content will require more policy attention in the future.

- Bottlenecks in the use of advanced mobile (wireless) broadband services and associated content should be resolved. Efforts are needed to move to more complex and data-rich mobile applications. Governments should assess how current market structures, competition, the affordability of mobile broadband access, and the lack of standards affect advances in this field. The access of new market entrants should be facilitated. Governments can also lead the way and promote increased mobile public-sector content usage such as health information, educational materials and other government-provided digital content.

- Governments have to renew efforts to put government services and government content online. E-government services and broadband applications would help organise the public sector more efficiently (also in areas such as public safety), however, these have not been developed sufficiently, even in leading OECD countries.

- Governments should move beyond plans to create access to and commercial use of public sector content information (essentially data), towards creating access to public and cultural content (*e.g.* museums). Putting the legal and technical infrastructure in place to make this happen, to allow for cross-border access and interoperability while avoiding the risk of information decay, however, will require sizeable efforts.

- It is crucial that government and business support the evolution towards more advanced broadband applications in social sectors such as telework, education, energy, health, and transport, where real progress is needed. Pressing societal challenges (*e.g.* pollution, ageing) persist for which effective broadband services could provide important solutions:

  - Despite early promises, these services and applications often remain in their infancy. Pilot projects need to obtain sufficient scope and scale and industry involvement in order to achieve critical mass.

  - Given the complexity of this undertaking, and considering the central role of governments in fields such as education, health and transport, a more active and swift approach is needed at this stage. Learning from existing public-private partnerships in this field across the OECD, sharing good practises and even co-operating with OECD member countries should be high on the list of policy priorities. The *2008 OECD Ministerial on the Future of the Internet*

Economy will aim at fostering these developments and raising these issues with Ministers.

Promoting competition, innovation, interoperability and choice

* For these increasingly complex broadband application markets, governments should intensify their efforts to promote competition, innovation, interoperability and choice.

* Maintaining a level playing field and reducing anti-competitive practices in the face of high network effects and to promote consumer choice is crucial, *i.e.* in particular considering the increased use of walled garden approaches, as well as cross-industry mergers and acquisitions. With problems such as vertical integration, lock-in of consumers in certain standards, and poor access to certain content, an environment of contestable markets should be created where small and innovative players can compete. Further analysis of recent trends and impacts of concentration is also needed. When necessary, anti-trust and other policies have the means to restore competition.

* It will be crucial to monitor and analyse the new market structures of broadband software, service and content providers in the next few years. Governments have a lot of experience when it comes to ensuring efficient telecommunications markets. However, when it comes to broadband applications, services, software and content, this is mostly new territory. It is important in the coming years that policy makers understand the impacts of new broadband market structures and question whether current policy approaches for ensuring competition actually work.

* OECD governments need to promote interoperability at the international level and encourage open standards. It is usually not up to governments to choose standards but they can play a role in encouraging and assisting industry co-operation (*e.g.* through setting up cross-industry fora on particular standards, or through engaging in the standard-setting process). Governments can mandate a certain degree of interoperability and promote open standards.

Security, privacy and consumer protection

* Ensuring the security of information systems and networks is vital. This must continue to be a policy priority in the years to come. In particular, governments' efforts in this area should be better co-ordinated at the international level, and should include increased law enforcement co-

operation. Computer security incident response teams should be improved, and there should be greater public education on security in general.

- Broadband uptake and Internet usage are growing. This raises privacy issues that need monitoring. Existing privacy policies need to be enforced and, where updated, to reflect new challenges.

- OECD countries should continue to develop more effective policies to protect consumers online.

- Policy makers, industry and civil society also need to examine new broadband consumer protection issues that are not currently addressed in the Recommendation. In particular, consumers can be confused by misleading messages about pricing and data tariff structures as well as the quality of broadband services provided (*e.g.* discrepancies between actual and advertised speeds, unreliable connections and limited customer support). Adequate and accurate information needs to be available so that consumers can make informed choices about service providers. They also need transparent low-cost procedures in place if they wish to change service providers.

- New consumer issues have emerged in other areas of broadband services and content (*e.g.* interoperability). Governments should discourage harmful business conduct and practices such as misleading advertising and unjustifiably long consumer lock-in periods. They should encourage greater transparency about the interoperability of different broadband services and content.

Regulatory frameworks that balance the interests of suppliers and users

- Balanced regulatory frameworks in areas such as intellectual property rights (IPR) will remain a top priority – even long after other goals such as basic broadband access have been achieved. Finding the right balance in this new environment and devising schemes that promote creativity and reduce piracy will take time.

- Many of the issues related to IPR and digital piracy will play out in the market place, in courts, and without government involvement. Government intervention is required when there is evidence that the market is not working or failing to evolve in a positive direction.

- OECD governments, however, are advised to continue monitoring developments closely and to adjust the regulatory system when necessary. Governments should encourage industry to find solutions to make rich content available over broadband networks. They may also act as facilitators of dialogue and consensus among different industry participants in the value chain. Governments should also foster the availability of public content on broadband networks.

- In addition, all stakeholders should periodically evaluate the need for greater international co-ordination and harmonisation of IPR-related matters. Regulations about technical protection measures and fair use will need reviewing, so that there is a necessary balance between content creation, innovation and fostering the participative web, as well as copyright enforcement.

- In this new technological environment, the possibilities offered by new forms of content creation and diffusion may – in certain cases – be best regulated through innovative policy approaches, provided there is evidence that existing approaches are leading to undesirable results. It is crucial that economic analysis underpins the proposed regulatory modifications in the area of IPR. It is also important that processes be open and that content creators and consumers are full stakeholders in this policy process.

R&D for the development of broadband

- Governments must intensify efforts to ensure there is sufficient R&D in the field of ICT, so that the economic, social and cultural effectiveness of broadband is guaranteed. The role of government and business in basic R&D may have to be reaffirmed. Any government neglect in this area should be monitored as well as examples of inadequate policy co-ordination, with the aim of increasing the efficiency of broadband-related R&D.

- The adequacy, effectiveness and appropriateness of existing government R&D support schemes (*e.g.* tax credits) and their role for broadband networks, services and content should be reassessed.

- Strengthening broadband research networks (grids), and facilitating international co-operation through such networks and collaborative research should be a policy priority.

- Plans to provide digital access to scientific information and research should be accelerated.

Evaluation and policy co-ordination

- OECD governments have to implement specific broadband policy assessment and evaluation procedures in order to more effectively appraise progress in achieving the goals of broadband policy.

- One clear need emerging from the monitoring exercise is for more harmonised data on broadband coverage, on actual speeds, prices and competition. Certain key indicators are not available to users, such as the actual broadband line speeds, data on how subscribers use their connections, and measures of mobile data access. This could be addressed by the OECD.

- Improved policy co-ordination among various agencies, ministries and the private sector will be essential. This is especially needed with advanced broadband applications in vital sectors such as health, transport and others areas where responsibilities are shared.

- International fora of exchange such as within the OECD should be fostered, so that good practices may be shared, and difficulties encountered may be resolved.

Introduction

In February 2004, the OECD Council adopted its Recommendation on Broadband Development (Annex A). The Recommendation calls on member countries to implement a number of policy principles to assist the expansion of broadband markets, to promote efficient and innovative supply arrangements, and to encourage effective use of broadband services. The principles focus on the "virtuous circle" between supply and demand.

The OECD Council has asked the Directorate for Science, Technology and Industry to monitor the developments in broadband in OECD countries in 2008 and this report is a response to that request.

Objective and structure

This report monitors broadband developments in OECD countries in four parts:

Chapter 1: Infrastructure

Chapter 2: Diffusion and usage

Chapter 3: Framework conditions: security, privacy and consumer protection, balanced regulatory frameworks, research and development

Chapter 4: Policy assessment and evaluation.

Each of the four chapters is broken into two subsections: one looks at trends and developments in the particular area, and the other assesses the application of the Recommendation. Annex B presents links to the broadband policies of OECD countries.

Chapter 1

BROADBAND INFRASTRUCTURE

Infrastructure: Market trends and developments since 2003

The growing economic and social importance of broadband has resulted in most member countries, as well as the OECD, monitoring markets on a regular basis. In particular, regulators have monitored broadband subscriber data in order to assess market penetration rates. The OECD has collected comparative data on broadband penetration on a quarterly basis since 2001. Broadband penetration is a significant indicator that allows countries to gauge their relative performance.

Figure 1.1. Five criteria for evaluating broadband markets

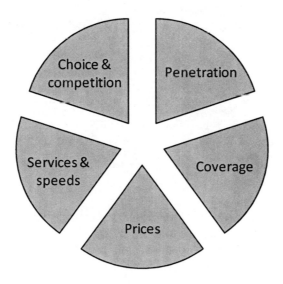

However, like any single statistic, it does not provide a complete picture of market developments. Figure 1.1 identifies five specific criteria used to evaluate the development of broadband infrastructure markets. Social and economic factors, while not included in Figure 1.1, should also be considered. The following sections will examine each in detail.

Box 1.1. Broadband subscriptions vs. broadband usage

Broadband has become a political issue in a number of OECD countries and penetration statistics are commonly used, or misused in debates and policy discussions. One of the most common sources of confusion is the fundamental difference between broadband subscriptions and broadband usage.

Subscriptions: Broadband subscriber data represent the number of physical connections supplied to subscribers by telecommunication operators. Regulatory agencies typically gather subscriber numbers directly from operators and then pass them on to the OECD. The benefit of subscriber data is that it is timely and provides an accurate tally of categories of broadband lines in service in a country. One drawback is that subscriber data cannot provide information on how any one line is being used, either in a household or a business. For example, a subscription to a home will only be counted once, even though five people living in the household may use it. The subscriber data only counts the number of actual subscriptions.

Usage: Broadband usage data are very different because they typically come from surveys/questionnaires given to a sample subset of the population. The results are then extrapolated for the country as a whole. National statistical agencies typically do surveys every one or two years because of the work and expense required. The benefit of usage data is that it provides detail on how many people use a connection and what they do once they are online. Survey data is typically reported in terms of the number and percentage of people or households *using* broadband. The drawback of survey data is that it is collected infrequently and the questions asked about broadband usage are not necessarily uniform across OECD countries.

Section one of this paper focuses on broadband *penetration* statistics – representing the number of physical connections provided by operators. Section two looks at *usage* data – from surveys – to gauge how households and businesses are actually using the connections.

Penetration

One of the reasons that broadband penetration levels have become an important litmus test for the state of broadband markets is because prices, coverage and competition levels are all factors in determining subscriber take-up. Economists have struggled with empirical tests which identify the determinants of broadband supply versus broadband demand due to a lack of data. As a result, member governments commonly use broadband penetration rates as a gauge of market development.[1]

In 2004, the year of the Recommendation, broadband was at an important stage of development: it was the year when the number of broadband subscribers in the OECD surpassed the number of dial-up subscribers. The shift to broadband thus led to a precipitous decline in dial-up subscribers (see Figure 1.2).

Figure 1.2. Growth of dial-up and broadband Internet access in the OECD, 1999-2005

Millions of subscribers

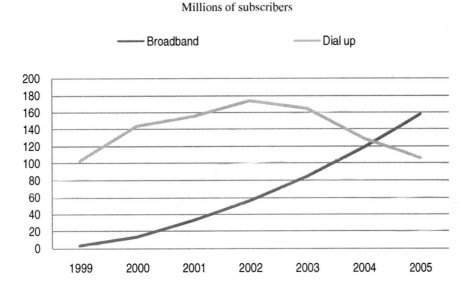

In the period 2003 to 2005, the shift away from analogue and early digital (ISDN) fixed lines was offset by strong increases in the number of higher-speed, broadband connections. Broadband connections overall grew by 88% over the period, with DSL and "other" technologies, dominated by fibre, showing the most rapid percentage increase. There were 83 million broadband subscribers in the OECD at the end of 2003. By June 2007, subscriptions had grown by 165% to 221 million (see Figure 1.3).

Figure 1.3. OECD total broadband growth, 2003-2007

In millions of subscribers

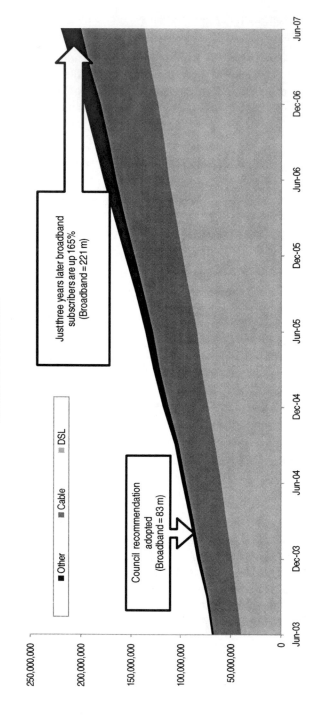

Figure 1.4. Broadband penetration, June 2007

Broadband subscribers per 100 inhabitants, by technology

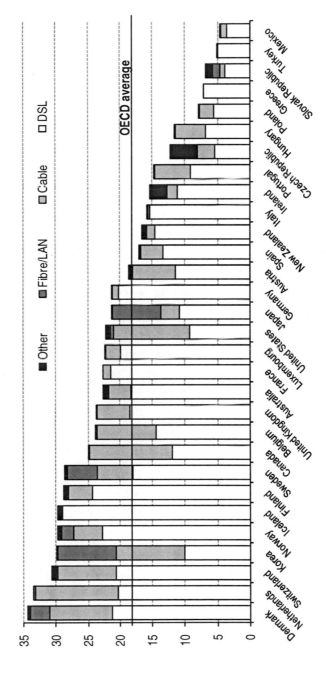

Broadband development in many OECD countries is growing; new connections are still increasing, however, at a slower pace. Penetration data may not capture the number of subscribers moving to faster connections (quality improvements in economic terms). While the growth rate is slowing down there will still be considerable internal changes as users upgrade their connections.

In June 2007, eight countries (Denmark, the Netherlands, Switzerland, Korea, Norway, Iceland, Finland and Sweden) led the OECD in broadband penetration, each with at least 28 subscribers per 100 inhabitants (Figure 1.4). Three countries (Denmark, the Netherlands and Switzerland) surpassed the level of 30 subscribers per 100 inhabitants.

Countries with relatively high broadband penetration rates tend to have relatively well developed communications infrastructure, as exhibited by fixed-line penetration rates and/or the number of households passed by cable TV. A number of other factors are certainly important in determining penetration rates but the correlation between GDP per capita and broadband penetration is relatively high.

Figure 1.5 shows broadband penetration across the OECD in June 2007 and GDP per capita from one year earlier. Holding all other factors constant, countries below the line have fewer broadband subscribers than their GDP alone would predict. On the other hand, countries with data points above the line have broadband levels which are higher than their GDP would suggest. The correlation between GDP and penetration is also relatively high (0.60) and captures likely many of the same factors as the relationship between fixed-lines and broadband (*e.g.* GDP and fixed lines are collinear). It is important to note that GDP is correlated with penetration, although the relationship is not necessarily causal.

Benchmarking broadband penetration allows policy makers to obtain a better picture of the relative performance of their markets both across countries as well as over time. In isolation, there is a tendency to view national growth rates as high, since, as with any new technology, growth is often in the double digits. A relatively slower broadband penetration rate may be indicative of certain market deficiencies, or may reflect other market or country-specific factors. It can be helpful for policy makers to compare broadband penetration levels with countries of similar income levels. This allows for a better comparison in many cases, and observers can highlight outstanding performances for a given income group.

Figure 1.5. Broadband penetration and GDP per capita

Subscribers per 100 inhabitants (June 2007) and GDP per capita (2006, USD PPP)

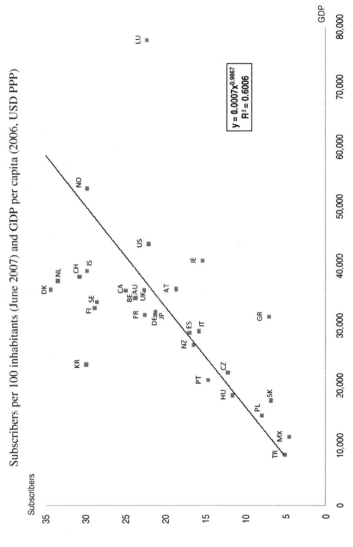

$$y = 0.0007x^{0.9867}$$
$$R^2 = 0.6006$$

Another important way to examine the growth of broadband in the previous three years is to compare broadband penetration levels in 2003 with 2006, that is, in the three years following the OECD Council Recommendation on Broadband Development. Figure 1.6 shows the broadband penetration rate for December 2006 but distinguishes between the level of penetration achieved before the Recommendation (light part of the line on the left) and any gains after the Recommendation (darker part of the line on the right).

Figure 1.6. Broadband penetration growth

Subscribers per 100 inhabitants growth (2000-2003, left) and (2004-2006, right)

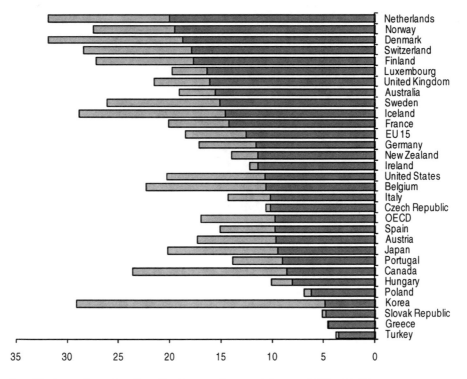

Note: Sorted in descending order of broadband penetration growth between 2004 and 2006.

There are several interesting elements that can be gleaned from the figure. First, Korea had achieved the vast majority of its penetration level before 2003. A combination of policy, geography and competition put Korea far ahead of its OECD peers in 2003 when the Recommendation was under formulation. In the three years following, other countries caught up, and in a few instances, surpassed Korea. It is worth noting that there have been significant gains in Korea since 2003, as users have migrated to faster fibre-based connections from DSL which are not captured in this particular indicator.

The largest growth in broadband penetration in per capita terms was in the Netherlands, Norway, Denmark and Finland – countries that now lead the OECD in penetration overall. The Netherlands added over 20 subscribers per 100 inhabitants in a three-year period to its total broadband subscribership, helping push it to the top two countries of the OECD, alongside Denmark. The penetration rate increase in the Netherlands over those three years, by itself, was higher than the *total* penetration of 16 OECD countries in December 2006. Policy makers should focus attention on what has helped propel these leading countries over the previous three years.

Clearly the blossoming of competition among providers in the Netherlands and Denmark has been a key factor in their strong penetration gains during the period and may also explain their leading places in the OECD as a whole. Both the Netherlands and Denmark benefit from infrastructure-based competition and same-line competition over DSL. In addition, fibre-to-the-home networks are appearing in both countries, often with the partnership of local municipality or utility company.

Coverage

Geographic conditions (*e.g.* total land mass, geographic dispersion of the population and terrain) can certainly increase the cost of initially connecting areas with high-speed Internet access. For example, providing backbone infrastructure between major cities and remote areas can be more difficult in geographically large countries with dispersed populations such as Australia. It may be easier for operators in countries with small land areas or with heavily concentrated urban areas to connect all users and these countries are often some of the first to reach high coverage levels with new technologies. For example, the penetration of fibre to the building and fibre to the curb has already reached 80% in Luxembourg, the smallest country in the OECD.[2] Difficult terrain can also be a factor, even in smaller countries such as Switzerland, which have extensive mountain ranges with remote villages.

Despite these challenges, a number of large OECD countries do have extensive broadband networks. In the OECD's largest country, Canada, virtually all households in urban centres and 78% of households in rural areas were within the broadband footprint at the end of 2006.[3] Broadband coverage is also extensive in the OECD's second largest country, the United States in areas already covered by cable or telephone service. High-speed cable modem service is available to 96% of end-user premises in the United

States where the cable systems offer cable television. In addition, DSL service is available to 79% of end-user premises in the United States where the incumbent local exchange carrier offers local telephone services.[4]

One explanation of why, in most instances, broadband penetration and a range of available geographic variables show little or no correlation is that large countries tend to have extensive coverage of DSL and cable networks. In fact, the total landmass of a country has a very low correlation with broadband penetration per 100 inhabitants across the OECD (see Figure 1.7). For example, Canada has the highest penetration rate among the G7 countries – which are all smaller.

Figure 1.7. Broadband penetration and total landmass, June 2007

Landmass (sq km) and broadband subscribers per 100 inhabitants

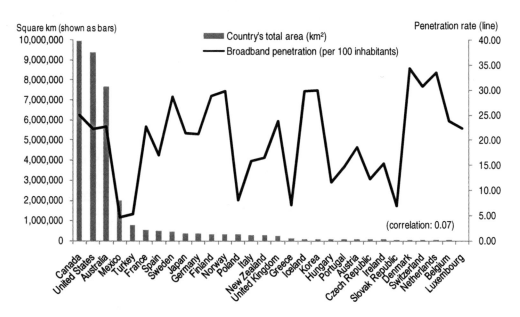

Source: OECD (broadband), FCC (landmass).

Another way to examine the effects of geography on penetration is by looking at the distribution of the population within countries. Theory would suggest that a country with the majority of its population clustered in a small geographic area should be easier to cover with telecommunication networks. This, in turn, could indicate higher potential for broadband penetration than a country with a more dispersed population.

Figure 1.8. Broadband penetration and population dispersion, June 2007

Population dispersion as measured by the percentage of the landmass inhabited by 50% of the population

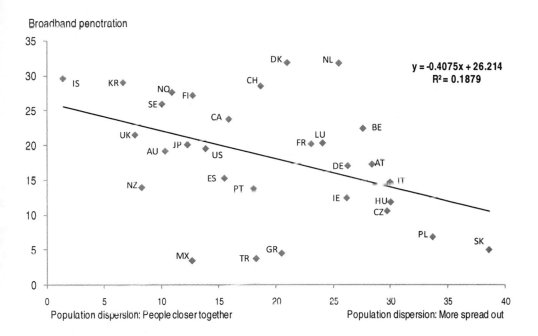

Note: Higher points on a vertical line represent higher penetration for a given density level.

Source: Broadband penetration (OECD), population dispersion (FCC, US).

Iceland, in particular, is an interesting example because the analysis changes drastically depending on which measure of geography is used. Iceland has the second lowest population density in the OECD when measured by the average number of inhabitants per square kilometre of landmass. An analysis using this data could conclude that connecting Icelanders might be more difficult than in other OECD countries. However, examining the dispersion of Iceland's population reflects a very different situation. Iceland has the least geographically dispersed population in the OECD, with 50% of its inhabitants concentrated in just 1.4% of its landmass. Basing an analysis on population dispersion instead of average density may show that connecting a large percentage of the population would only require a rollout in a relatively small geographic area. Iceland has the lowest population dispersion in the OECD, while the Slovak Republic has the highest. Interestingly, the correlations of broadband penetration with average density per square kilometre and with population dispersion (Figure 1.8) are both insignificant.

Geography can still be a factor affecting penetration rates, despite the low correlations found in the data. Most OECD broadband subscribers receive their service from networks that were built before the development of DSL and cable modems. Operators had to upgrade equipment to provide broadband service over the existing lines, but the underlying, extensive last-mile networks were already largely in place. In this case, the level of competition, prices and a number of social variables are likely to be more important determinants of broadband take-up (over current technologies) than broadband coverage. However, geography will become an important element in the deployment and take-up of new fibre-based, next-generation networks. Large countries with dispersed populations living away from city centres may initially have lower penetration levels than their peers due to difficulties in economically expanding next-generation services to rural and remote areas. Geographical conditions will also influence the decisions of operators deciding whether or not projects are economically feasible.

Networks expanded

Large businesses in central business areas and urban dwellings were well covered by operators networks in 2003 but many smaller businesses and households were unable to access services since incumbent telecommunication operators had not upgraded their networks to support DSL. Over the past three years, network operators made impressive gains upgrading exchanges and cable nodes to reach these small businesses and residential users.

Broadband coverage in 2005 was nearly 90% or more in a number of OECD countries over at least one physical network (typically with DSL), with many subscribers able to choose between multiple providers (see Figure 1.9).

Figure 1.9. Infrastructure-based broadband competition in selected OECD countries, 2005

Percentage of population covered by wired broadband technologies (DSL and cable),

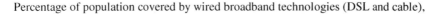

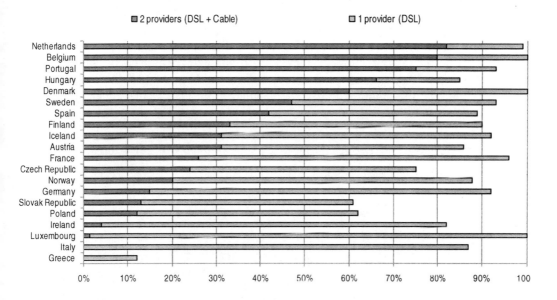

Source: OECD graphic based on data supplied by the European Commission.[5]

In the OECD area, DSL networks have the most extensive broadband coverage overall. DSL coverage is particularly high in Belgium, Korea, Luxembourg, the Netherlands and the United Kingdom. In 2005, 22 OECD countries had at least 90% coverage measured by lines, households or population. Greece had the lowest DSL coverage in the OECD area with only 9% of the population able to obtain a DSL line in 2005 (see Figure 1.10).

Cable providers have made impressive gains upgrading networks and offering broadband services to the majority of homes previously without cable television. Broadband coverage by cable networks is very high in countries such as the United States, Canada, Korea, Belgium and the Netherlands. In some areas it is even more extensive than DSL. In 2005, a number of countries had very high percentages of television viewers who could have access to television signals via cable, such as Belgium (88%), Korea (77%), the Netherlands (92.3%) and Switzerland (89.9%) and the United States (99%).[6] High levels of cable television coverage also correspond to high levels of cable Internet coverage. In countries such as Greece,

Iceland and Italy there is very little or no availability of cable networks. In other markets, cable coverage may be limited only to large metropolitan areas.

Almost all broadband connections in the OECD are now over DSL and cable lines. DSL accounts for the largest portion (62%) with cable subscribers numbering just about half of DSL (29%). There are other technologies that have increased in importance over the last few years. Operators installing new wired networks are increasingly using fibre optics instead of copper. Fibre networks are preferred in new infrastructure developments because the public works component is roughly 70% of the total cost of the network rollout, and the additional costs of installing fibre instead of copper in the ducts are minimal.[7]

Figure 1.10. DSL coverage and population density, 2004-2005

DSL coverage (% upgraded lines/population/households) and inhabitants per square kilometre

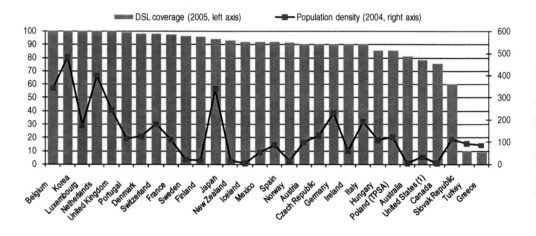

Note: DSL coverage is measured in various ways across the OECD. The percentages given above may represent the number of lines that have been upgraded, the population covered or the households which are able to subscribe.

(1) Data for the United States is an average for Verizon, SBC, Bell South, Qwest, Sprint, Alltel, Cincinnati Bell, Centurytel and ACS.

Fibre networks are preferred because the capacity of fibre is much higher than traditional copper lines and capacity is relatively easy to expand once the fibre is in place simply by adding additional lasers to a line. Fibre-to-the-home networks are expanding in countries across the OECD and in other parts of the world (see Figure 1.11). Many of these networks have been in metropolitan areas since the density reduces infrastructure costs on a per-subscriber basis. The cities of Amsterdam, Vienna, Reykjavik and Paris all have FTTH networks in the planning or rollout stages. Telecommunication operators themselves are moving to FTTH rollouts in various countries as well. NTT of Japan has the largest FTTH network rollout in the world in terms of total homes connected. Verizon in the United States is upgrading users to fibre connections and they plan on passing 9 million homes with fibre by year-end 2008 and 18-20 million homes by 2010.

Figure 1.11. Economies with more than 1% fibre-to-the-home/building penetration, July 2007

Percentage of households

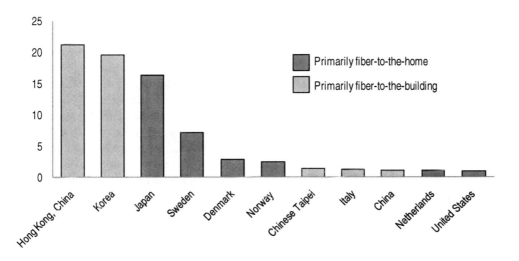

Source: FTTH Council.

Interestingly, some of the key developments in fibre deployments have been away from the main city centres. The previous three years have seen a surge in the number of smaller communities investing in fibre-to-the-home infrastructure. For example, the town of Nuenen in the Netherlands with 8 000 homes is reported to have passed 7 200 in the town and signed up 6 500 subscribers for FTTH services.[8] Communities in the Netherlands and Northern Europe have shown interest in helping build fibre networks but there has been activity across the world from Vienna[9] to Vermont.[10]

Denmark leads the OECD in broadband penetration rates and fibre rollouts have become a key component of Danish broadband access. What sets Denmark apart from other OECD countries is the participation of local power companies in delivering fibre-to-the-home access. Traditionally electric companies have been seen as potential broadband competitors using power-line communication technologies. However, some operators have found that fibre-based technologies can be installed as a way to reduce costs on the electricity side of the business (*e.g.* through automated metre reading) while increasing revenues through providing data services to end-users. Utility companies also have existing "rights-of-way" that simplify some aspects of network rollout.

Wired connections offer the fastest connections and the lowest prices per Mbit/s in the OECD when they are available. However, there are still many communities in the OECD that do not have access to wired broadband infrastructure. As Internet demand in rural communities has grown, there have been a variety of developments in wireless broadband.

Fixed wireless access has become available in some rural areas but these networks serve only a small percentage of subscribers. The exception would be operators in the Czech and Slovak Republics, for example, who use a range of fixed-wireless technologies to reach households. In the Czech Republic alone, fixed wireless providers using technologies such as CDMA 450 accounted for roughly 34% of all broadband connections in the country. There has been debate as to whether these connections constitute "broadband" under the OECD definition, but the reported connections should only include fixed wireless access subscribers with speeds greater than 256 kbit/s. In countries such as the Slovak Republic, wireless broadband could be very important since a smaller percentage of subscribers live within range of an upgraded DSL exchange than most of the OECD.

Wireless Internet access depends on available spectrum and OECD countries have taken steps to improve the efficiency of spectrum use. The National Telecommunications & Information Administration (NTIA) in the United States worked with federal users to free up radio frequencies for commercial and other uses such as public safety. The United States Federal Communications Commission (FCC) has also been working to make a significant amount of spectrum available for wireless broadband services. In September 2006, the FCC completed its auction of 90 megahertz of Advanced Wireless Services spectrum. Then in January 2008, the FCC began auctioning an additional 62 megahertz of spectrum in the 700 MHz band, which is particularly well suited for wireless broadband.

Internet access via mobile phone or laptop computer connected to a 3G network (if at broadband speeds) would theoretically have the potential to reach a much larger number of subscribers than wired broadband, even in the most advanced broadband countries. In fact, the number of 3G subscribers with broadband-speed mobile connections in Korea far surpasses the total number of Korea's broadband connections reported by the OECD. Korea had 14 million broadband subscribers using the OECD methodology in 2006 but 36 million 3G wireless subscribers. Combining the two would create a fixed/portable broadband penetration rate of 103 subscribers per 100 inhabitants.

Because of their reach, wireless Internet connections using 3G or emerging wireless networks will be an increasingly important but largely complementary access technology to wired broadband. OECD countries already have extensive 2G coverage and many of these networks are likely to be upgraded to 3G in the near future. All OECD countries have 2G mobile coverage of more than 90% of their populations.[11] Even large countries with extensive rural areas typically have excellent coverage of places where people live. Data shows that subscribers are switching to 3G networks nearly as rapidly as they originally took up cellular/mobile phones. Third-generation mobile data coverage is very high in a number of countries including Sweden, Korea, Luxembourg, Italy, the United Kingdom and the United States.

It is important to note that 3G subscribers are not necessarily using their phones to access the Internet and that generally mobile access to the Internet and the use of mobile applications is lagging in all but a few countries (see the use section for more information).

The broadband technology with the broadest geographic coverage is satellite. Geo-stationary satellites can supply broadband over very large geographic areas. Early satellite broadband connections required a fixed-line return path (upstream data) but current terminals can now transmit and receive data.[12] Satellite has a large coverage area but only accounts for a small fraction of OECD broadband connections – largely due to its relatively high price compared with other connectivity options. Satellite connections are used for backhaul and end-user connections in rural and remote areas and play a vital role connecting areas that have no other means of access.

Growth has been uneven

The coverage of wired and wireless broadband technologies described above has grown over the previous three years but this growth has been uneven in some respects. There has been significant progress in upgrading telephone and cable television networks with broadband capabilities, but most of this development has centred on urban areas. Many rural and remote areas are still waiting for high-speed broadband connectivity.

The Recommendation makes particular mention of rural and remote areas. It highlights broadband affordability in rural and remote regions and states that governments need to encourage broad geographic coverage, particularly in these under-served areas.

Nearly all countries have developed strategies and programmes to expand coverage in rural areas. One example was Canada's Broadband for Rural and Northern Development Pilot Program which was created as a way to bring connectivity to "First Nations, northern and rural communities". The project funded 154 projects representing roughly 2 285 communities.[13] The BRAND Pilot concluded on 30 March 2007.

Digital divide issues were an important subject of a 2004 OECD conference on developing broadband access in rural areas. Three years later there has been significant progress towards reaching rural and remote areas with broadband, however, promising new technologies such as WiMAX and power-line communication (PLC) have not been significant in these broadband deployments. Cable, DSL, satellite and various fixed-wireless access technologies are still the technologies supplying rural areas. In particular, new DSL technologies that increase the distance of data transmission have played an important role.

WiMAX was initially promoted as a key technology for rural areas. Clearwire, in the United States, launched wireless broadband service in 50 smaller (tier three) markets in 16 states across the US. However, the largest WiMAX deployments are in metropolitan areas well covered by fixed-line broadband connections. Network operators are choosing the mobile WiMAX standard (IEEE 802.16e) for dense urban environments rather than deploying the fixed standard (IEEE 802.16a) that is capable of longer distances. For example, Korean operators launched mobile WiMAX (called WiBro in Korea) in Seoul in 2006.[14] There are operators who are using WiMAX in rural areas. Policy makers are also looking at a number of emerging wireless technologies to connect users at very high speeds. These include long-term evolution (LTE) or '4G' mobile networks.

Another technology receiving a good deal of attention in 2004 was power-line communication (PLC) but the number of actual PLC subscribers remains very low. As in 2004, discussions surrounding PLC are upbeat in 2007 but there has still been very little deployment to date. Denmark had 98 PLC subscribers at the end of 2006, while the United States had just over 5 000 in June of the same year. The technology may still be promising, particularly for rural areas, but it has not made an impact on broadband penetration over the past three years.

Broadband users in rural areas also face increasingly disadvantaged access to bandwidth in relation to their urban peers. In the early days of broadband, a fast Internet connection was roughly five times as fast as dial-up connections available in rural areas. Speeds have increased for almost all broadband subscribers as technologies improve but not for those left with dial-up connections. The divide between rural and urban areas continues to grow.

Figure 1.12. OECD broadband and dial-up speed comparisons, 2004 and 2007

In advertised kilobits per second

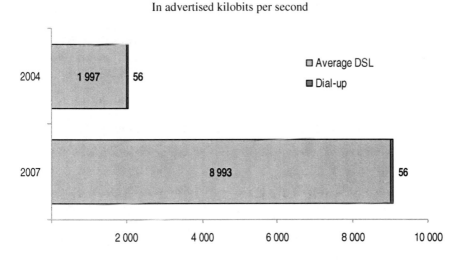

In 2005, dial-up still accounted for 40% of all fixed Internet connections. Figure 1.12 highlights the growing disparity between Internet speeds available in urban areas and the dial-up connections still common in rural and remote areas. In 2004, the average advertised DSL speed in the OECD was 36 times faster than a standard dial-up connection. However, by 2006 the average DSL connection was 160 times faster than a standard dial-up connection. This has created problems for those left behind still connecting via dial-up modems since content providers are increasingly building sites and services geared toward the faster connections available in urban areas. This will be discussed further in Chapter 2.

Prices

Prices have an impact in areas with wired coverage as well and can be a strong determinant of broadband take-up. The prices of broadband services vary widely across the OECD and are commonly lower in markets with high levels of competition. Comparing prices can be difficult because speeds, quality and levels of mobility vary for many different broadband technologies.

Broadband is affordable in most OECD countries. The average monthly price of a broadband subscription in 20 of the 30 OECD countries was less than 2% of monthly GDP per capita in October 2007 (see Figure 1.13). It is worth noting that subscribers in rural and remote areas may face higher prices. The lowest-priced, entry-level broadband plan in all countries was never more than 3% of monthly GDP per capita and was typically less than 1%.

A sample of similar broadband plans across the OECD in September 2005 and October 2006 shows that prices for the same (or a slightly improved) service fell on average 19% for DSL and 16% for cable.[15] This has allowed more subscribers from a much wider range of income groups to access broadband than in 2003. Coverage has increased and more people within the expanded footprint can afford services.

Figure 1.13. Broadband affordability, October 2007

Entry and average monthly broadband price as a percentage of monthly GDP per capita

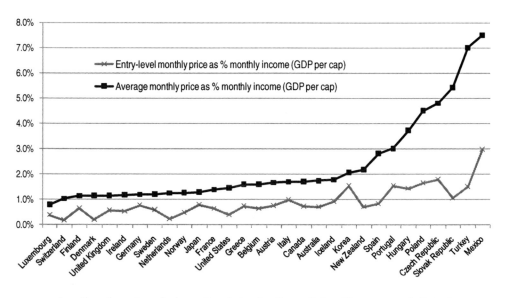

Another key trend since the introduction of the Recommendation has been the development of "entry-level" and "high-end" subscriptions as broadband operators expand the number of different offers they make available to consumers. It is worth noting that some higher-priced plans in

countries such as Spain were discontinued after the most recent collection of data in October 2007. The introduction of entry-level broadband plans has helped broadband operators increase the number of total subscribers while still offering the possibility for customers to pay more for faster speeds.

These differentiated service levels have led to a range of broadband prices in each country. Entry level plans now start as low as USD 5.80 (PPP) per month in Switzerland and can go up to over USD 322 (PPP) per month for a connection in the Czech Republic (see Figure 1.14). Interestingly, some of the leading countries in OECD broadband penetration also have the smallest price differences between their entry-level and highest-priced broadband plans.

Broadband has also reduced prices for several related telecommunication services such as voice telephony. There has been a shift over the past three years away from paying for individual voice calls to flat rate voice calling plans. In many cases this has reduced the prices that households pay for these services when bundled together.

Unbundling of copper telephone lines itself seems to be a factor in reducing the price of broadband subscriptions, as they introduce more competition at the telecommunication exchange. Evidence points to lower 'per Mbit/s' charges in countries with unbundling rules. Prices per Mbit/s were significantly higher in the least expensive of the four countries, with limited or no unbundling compared with other leading broadband economies. The price per Mbit/s in Japan was USD 0.22 per Mbit/s while the least expensive Mbit/s in the United States was 14 times more expensive (see Figure 1.15). Switzerland adopted local loop unbundling in 2007 and is among the top five OECD countries in terms of penetration. However, in 2006 *before unbundling*, Switzerland's price per Mbit/s, even in PPP terms, was 19 times more expensive than in Japan and five times more expensive than in neighbouring France.[16]

Speeds and services

Competition is a key to lowering prices but it also has a significant effect on the services and speeds available to businesses and consumers. Broadband quality tends to increase over time even as prices decline. This is a common feature in the ICT sector but broadband changes have been particularly rapid. At the end of 2004 the average DSL speed across the OECD was less than 2 Mbit/s. The average advertised broadband speed had more than quadrupled to nearly 9 Mbit/s over a period of less than three years.[17] The trend continues as operators upgrade their networks.

Figure 1.14. Broadband monthly subscription prices, October 2007

All platforms, USD PPP, logarithmic scale

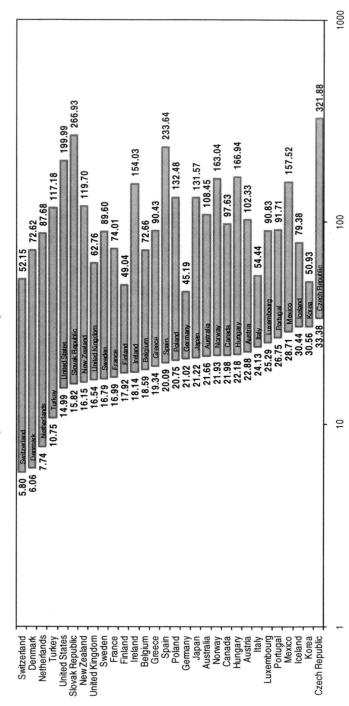

Figure 1.15. Broadband prices per megabit per second, October 2007

All platforms, USD PPP, logarithmic scale

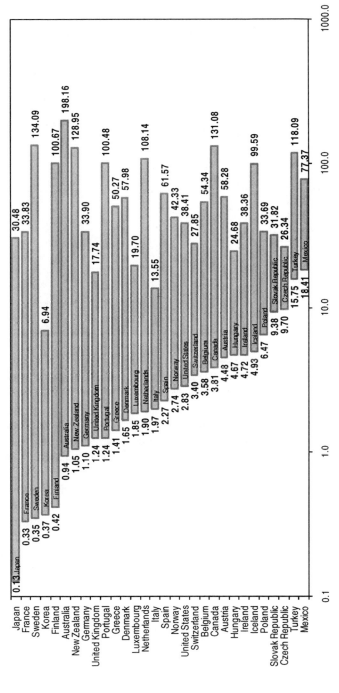

Figure 1.16. Fastest broadband download speeds offered by the incumbent telecommunications operator, October 2007

All technologies, megabits per second

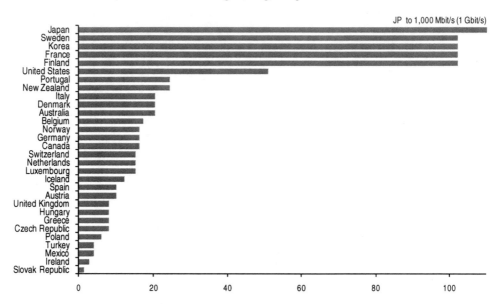

Note: The connections represented are over DSL, cable or fibre and they refer to the fastest consumer speed available in October 2007 from the incumbent operator on the date the data was gathered. The top speed plan in the United States is from Verizon.

As mentioned earlier there are still significant speed differences between rural and urban areas and this has led to some concern that a significant digital divide will continue to exist.[18] Further research is needed on the breakdown of speeds that consumers choose to buy since this could help reveal the demand for bandwidth and the extent to which rural areas may be disadvantaged. The European Commission finds that download speeds between 144 kbit/s and 512 kbit/s have been the most common in rural areas in the past two years. In contrast, the most common speeds in urban areas are closer to 1 000 kbit/s. They also found that there is a clear trend towards higher speeds in urban areas, while speeds in rural areas tend to remain constant.[19]

Cable Internet connections were typically faster than DSL connections in the same country in 2004. However, average advertised DSL speeds have grown more quickly and this is likely the result of fewer technological upgrades on cable networks in comparison to DSL providers, many of whom upgraded to faster DSL specifications over the previous three years.

By October 2007, the fastest advertised broadband connections offered by incumbent telecommunication operators were in Japan, Korea, Sweden, France and Finland. NTT in Japan offers 1 Gbit/s connections to apartment buildings (1 000 Mbit/s) while the other operators offer FTTH at 100 Mbit/s to individual apartments or houses (Figure 1.16).

The cable operators with the fastest advertised speeds over their cable infrastructure are Australia, France, Japan and Luxembourg (see Figure 1.17). The French cable operator Numericable has begun offering 100 Mbit/s connections but only in areas where it has upgraded with fibre. The operators in all four countries advertised broadband at speeds of 30 Mbit/s or greater. Cable providers in 17 countries advertised broadband at speeds of 10 Mbit/s or greater. Only 13 incumbent DSL providers in the OECD advertised similar speeds. Interestingly, cable top speeds in the two countries dominated by cable broadband subscriptions, Canada (25 Mbit/s) and the United States (20 Mbit/s), were slower than advertised offers in a number of other countries dominated by DSL.

Figure 1.17. Fastest broadband download speeds offered by surveyed cable operators, October 2007

In megabits per second

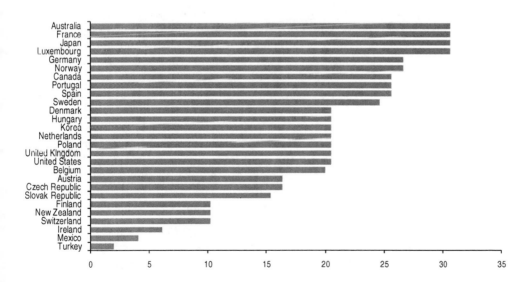

Broadband in the OECD is still dominated by DSL (63%) but there is an obvious trend emerging to upgrade last-mile, copper-based connections to fibre. The transition to high-definition television (HDTV) signals has been one of the main reasons operators have decided to expand capacity. The bandwidth required for a standard television signal over IP is roughly 2 Mbit/s, an amount that is typically advertised to DSL subscribers in the OECD. However, the amount of bandwidth needed for one HDTV television signal will be significantly higher, roughly 10 Mbit/s for each channel streamed (depending on compression techniques).

Korea and Japan have the most subscribers with access to the Internet via fibre-based connections and these offer the fastest broadband download speeds. Japan leads the world in fibre-to-the-home connections with 9.7 million subscribers in June 2007. Korea has both fibre-to-the-home and fibre-to-the-building (called 'apartment LAN') which can provide speeds of 50-100 Mbit/s. Korea had 4.5 million fibre-based subscribers in June 2007. The total number of DSL subscribers has fallen in both Korea and Japan as users upgrade to these fibre-based connections.

The speed of wireless Internet connections over mobile networks has been increasing as well. For example, 3G technologies, including forth-coming High Speed Packet Access (HSPA) are capable of a maximum bit rate of 14.4 Mbit/s per user on a cell. However, the UMTS Forum estimates that the typical throughput of one of these technologies, High Speed Downlink Packet Access (HSDPA) will be 500-700 kbit/s with a maximum of 40 users per cell.[20] As speeds increase, these wireless connections will become increasingly important access options in a number of OECD countries.

Advertised speeds

Offers for broadband typically advertise their maximum speeds, and this has been a source of confusion for many users over the last three years. Users subscribing to 8 or 24 Mbit/s plans are told the maximum speeds for ADSL (8 Mbit/s) or ADSL2+ (24 Mbit/s), even though their speeds ultimately depend on their distance from the exchange and a number of other factors.

A report from the United Kingdom regulator OFCOM found that only 20% of customers live close enough to a telephone exchange (3.2 kilometres) to receive 8 Mbit/s from an advertised 8 Mbit/s connection. If a user lives even further from the exchange, say 8 kilometres away, users may only receive between 0.5 and 2 Mbit/s.[21] Users may also expect connections to be capable of much higher throughput than is actually possible. This issue has emerged as users demand more bandwidth for their applications and services.

There have been some calls for operators to make more information available to consumers on the speeds that they are actually capable of receiving. Some operators have expressed concern that this information may be too technical for many consumers or that it would be too difficult to provide. However, a number of DSL operators in the OECD allow consumers to test a line and many provide detailed information about their line, information about the distance from the exchange, attenuation as well as estimations of maximum throughput.

Some governments have called for broadband advertisements to reflect the true bandwidth available to consumers. The Czech government's Broadband Plan proposes that the actual achieved rate (effective rate) over the long term should not be less than 80% of the nominal (advertised) rate.[22]

Contention ratios

As mentioned previously, distances between a DSL subscriber and the exchange partially determine the speed of subscriber connections. In addition, contention ratios (the measure of how many subscribers share a limited amount of bandwidth) have an impact on the speed of a connection that will affect all types of broadband subscribers. Consumers are increasingly interested in how much bandwidth they are ultimately sharing.

Telecommunication operators plan for network demand on the assumption that not all connections will be in use at the same time. For example, a provider may sell a 2 Mbit/s connection to 10 people in a neighbourhood. If the provider assumes that only 10% of subscribers connect at a given time then they assign the capacity "sold" to one subscriber (2 Mbit/s) to the entire group of 10 subscribers. Since only one person connects at any given moment they still receive the 2 Mbit/s they have purchased. However, if there are periods when multiple users are downloading at the same time, then the 2 Mbit/s must be shared among them all and actual throughput will fall for all subscribers.

This example is a simplified version of contention ratios but illustrates how the bandwidth users buy depends not only on distances from the exchange (with DSL) but also on the capacity of the node and the number of subscribers attached to it.

BT Wholesale in the UK provides two contention levels; 20:1 and 50:1.[23] This implies that either 20 or 50 subscribers share the bandwidth sold to each subscriber. A subscriber may have access to 8 Mbit/s when there is no other traffic, but speeds could theoretically fall to as low as 40 kbit/s if all subscribers attempted to download a very large file at the same time.

Concern over contention ratios has begun to emerge within the three years following the Recommendation: discussion on how bandwidth is distributed to end users will likely continue. Operators also need to be clearer about the contention ratios on the lines they sell.

Usage limitations (bit caps)

Another related trend to emerge over the past three years is the introduction of bit caps (data caps) in many OECD countries. As contention ratios have shown, broadband services were originally sold as unlimited services on the assumption that users would consume, on average, only a small portion of the bandwidth allocated to their connections. However, the development of various high-bandwidth applications such as peer-to-peer file sharing and video streaming have created more contention issues on broadband lines. Some operators responded by imposing limitations on the amount of bandwidth that users are allowed to transmit in a given month. These bit caps were typically found in island countries with limited international transmission capacity, but now they have appeared in other OECD countries as well. Currently there are offers with explicit bit caps in two-thirds of the OECD countries.

The amount of allotted traffic before reaching a bit cap varies from country to country. In October 2007 there were no explicit bit caps among offers surveyed in Finland, France, Germany, Italy, Japan, Korea, the Netherlands, Norway, Sweden and the United States – although heavy, continued usage could go against acceptable use policies and lead to the termination of an account. On the other hand, there were no surveyed offers *without* bit caps in Australia, Belgium, Canada and New Zealand.

Statistics New Zealand released a report in March 2007 that explained how bit caps were applied to users in the country. They found that among the 611 600 non-analogue Internet subscribers on 30 September 2006, only 2.4% had no data allowance cap (data cap) on their subscription plan. Additionally, 68.6% of subscribers used plans with a data cap of less than 5GB (419 600), 25.9% used plans with a data cap between 5GB and less than 20GB (158 700), and 3.1% had plans with a data cap of 20GB or more (18 700).[24]

Bit caps put limitations on how people can use their Internet connections. Policy makers in the Czech Republic, for example, have highlighted this effect and call for limitations of how much Internet traffic Czech consumers can use each month to disappear over time.[25] This may become an economic disadvantage in countries with relatively low bit caps, particularly as more high-bandwidth applications appear. On the other hand, some have argued that bit caps are an equitable way to ensure that users pay for the bandwidth they consume.

Competition

Having access to broadband and having access to a number of competitive broadband providers are quite different. As mentioned in the pricing section, broadband-type download speeds are available to almost any subscriber in the OECD via satellite but often at a very high price. The fastest connections, lowest prices and most innovative services are in areas where there is a range of consumer choices for broadband. One difficult issue for policy makers over the past three years has been determining which connections are substitutable for one another.

As an example, a 256 kbit/s connection is the bottom threshold for OECD broadband penetration statistics and is roughly five times faster than a dial-up connection. However, some ADSL subscribers are currently sold connections with theoretical line speeds up to 28 Mbit/s, more than 100 times faster than an entry-level broadband connection. Some argue that there are key differences in quality between a connection at 256 kbit/s and another at 100 000 kbit/s.

This has led to discussions on how to categorise broadband in the future. The level of infrastructure-based competition in broadband markets will be directly tied to how broad a definition policy makers take for "broadband". Markets relying on infrastructure-based competition alone have typically been limited to a maximum of two "wired" providers (DSL and cable). Wireless data options may also be available from satellite or mobile networks, although at speeds significantly lower than typical wired connections.

Infrastructure-based competition expanded over the past three years in the OECD as cable and DSL providers extended the reach of their data networks. Infrastructure-based competition from wireless Internet technologies has been limited over the previous three years as operators have been waiting on new technologies. A number of wireless technologies (including 3G) were touted as serious alternatives to DSL and cable connections, but per-kilobyte pricing plans on mobile networks in most OECD countries have made this an unreasonable assumption. Most wireless subscribers on flat-rate plans have been limited to using only low-bandwidth applications and still relying on wired connections for bandwidth-intensive services.

Network upgrades to faster mobile broadband technologies and flat rate pricing are now making mobile/wireless broadband connections more attractive to users. Mobile data networks in countries such as Australia help provide broadband in otherwise unserved areas and are beginning to replace ISDN services in areas without DSL or cable coverage. These networks may also be an appealing option for low traffic volume users in metropolitan areas.

The strongest new infrastructure-based competition has come from fibre networks built by new entrants who can attract consumers by offering speeds greater than cable or DSL. Competitive providers, municipalities and power companies have installed these networks. The areas where they have had the most significant positive impact on competition have been where networks have been run on "open access" principles. One example is the STOKAB fibre network in Stockholm, Sweden, which has approximately 60 telecommunication companies as clients serving subscribers using STOKAB's fibre lines.[26] Open access networks separate the provision of bandwidth from the delivery of services. Operators of open access networks typically allow any operator to sell their services over the network for a standard rate.

Competition issues over the future of unbundling have also come to the forefront of national debates.27 Operators in several countries have expressed interest in rolling out new VDSL networks that still make use of the copper network but from a much closer fibre node. VDSL connections use fibre-optic cables to reach a neighborhood street corner and then connect to homes using historical copper telephone lines for the last few hundred metres. This permits much higher speeds than are possible over ADSL. At the same time, VDSL deployments raise the issue of stranded investment of competitive equipment already located in exchanges in those countries where unbundling is mandated.

Debates on the future of unbundling have been heated, and these will likely continue. However, more recent debates have focused on the role of functional and structural separation of the local loop from the incumbent's retail activities. These are likely to be key issues in future policy debates. The reason the VDSL debates are so important is that a number of OECD countries have relied on unbundling for broadband competition.

While there has been significant progress with unbundling in many of these countries, more progress is needed to make markets fully competitive. Data from the European Competitive Telecommunications Association (ECTA) show that the number of exchanges hosting competitive equipment remains low in many OECD countries. This could have implications for the competitiveness of the broadband connections in OECD markets which rely on unbundling (Figure 1.18). This does not preclude unbundling, however, as unbundling rules in many OECD countries still allow subscribers to access broadband from competitive operators via wholesale arrangements.

Figure 1.18. DSL competition at the exchange, 2006

Main distribution frames hosting competitor's equipment, select countries

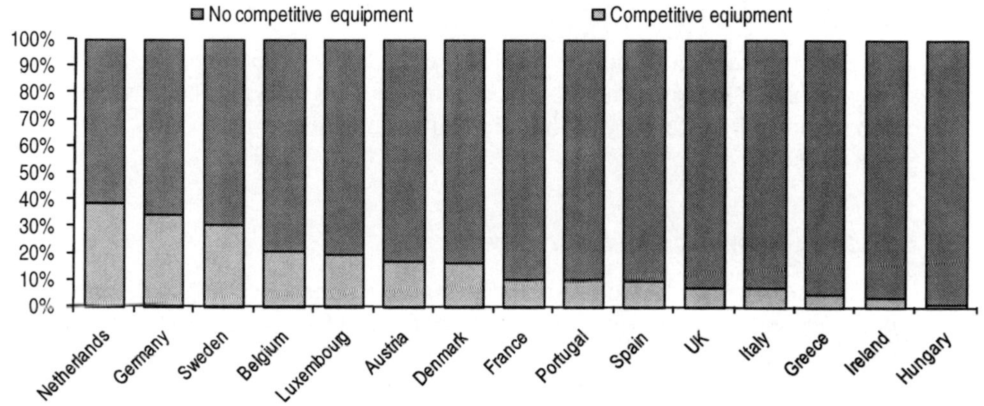

Source: ECTA.

Certain countries with infrastructure-based competition and unbundling rules have competition from not only the cable operator and incumbent telephone company, but also additional market players who rely on unbundling. This has reduced the danger of a duopoly market structure. As an example, the Netherlands has strong infrastructure-based competition between cable and DSL (see Figure 1.14) but also leads in ECTA's analysis of main distribution frames upgraded with competitive DSL equipment.

Countries such as the United States have chosen to focus on infrastructure-based competition in broadband markets due to the manner in which intermodal competition arose in the marketplace, as well as concerns that unbundling could create investment disincentives and limit the rollout of new networks. Policy makers in the United States are looking to power-line communications and wireless technologies as important competition to existing fixed line and cable providers.

Infrastructure: Application of the Recommendation

The previous section examines overall market developments in infrastructure since 2003 and provides an overview of general trends. This section applies a much closer focus – on specific policies and programmes put into place by member countries which help achieve the goals of the Recommendation. The topics related to infrastructure in the Recommendation fall into two broad categories: efficient markets and investment promotion. This section discusses all country-specific elements under one of these two headers.

Efficient markets

Telecommunication policy makers are always concerned with the efficiency and competitiveness of telecommunication markets. This focus on competition and efficiency is highlighted in the first three points of the Recommendation.

Effective competition and convergence

The first recommendation to emerge in the Council Recommendation is a need for regulators and policy makers to foster effective competition[28] as markets move towards convergence. The term "convergence" is somewhat vague and policy makers in many countries have interpreted it differently. The difficulty of defining "convergence" across OECD countries signals that there is still ambiguity among regulators about how markets will eventually evolve. It is vital in these situations that markets are efficient and competitive so they can be left to develop, relatively unhindered.

Business and residential customers benefit from increased competition. A report from an independent Canadian panel of experts, the Telecommunications Policy Review Panel (TPRP), highlights why competitive telecommunication markets serve consumers and the economy in general better than regulation and other government intervention.

"Setting prices and conditions of service that benefit both service providers and customers requires large amounts of information, more than a single organisation can easily gather, keep up-to-date and use. This is true whether the organisation is government or private sector. In competitive markets, changes to prices and conditions of services are generally made by trial and error, taking into account what has worked in the market and what has not. Competitive market forces can process more information and do so more efficiently than any single service provider or regulator."[29]

Competitive markets are better at managing the vast amounts of information that go into determining prices for telecommunication services. Policy makers in all OECD member countries have made stimulating competition in broadband markets a high priority. The means of promoting competition may vary from country to country but the goals are the same.

Transparency

The Recommendation calls for transparent and non-discriminatory market policies and similar phrases have been adopted in some OECD member countries' broadband policies. Markets characterised by large fixed costs and high entry barriers require transparent and clear regulations in order to promote investment.

One of the key ways governments have increased the transparency of the regulatory process is by opening up issues in front of the regulator to public consultation and making submissions a matter of public record. For example, regulators such as the Federal Communication Commission in the United States have given concerned market participants the opportunity to comment on and recommend action on issues before the Commission. The penetration of broadband to consumers and businesses has also indirectly helped increase the transparency of the process by allowing users to file comments electronically such as with the FCC's ECFS System.[30]

Recently, some policy makers and experts have put forward ways to improve transparency of the rulings once they have been announced. The Canadian expert group suggests that each newly established or amended rule should have, as part of the regulatory framework, a separate "rule" or "order" that could clearly describe the actual rules coming out of the decision. They also suggest that this document contain clear references to other applicable rules or decisions.[31]

Unbundling

Policy makers in 28 of the 30 OECD markets have adopted unbundling as a way to introduce competition into broadband markets. The decision to unbundle is usually seen as a "second-best" alternative to extensive infra-structure-based competition. Ofcom in the United Kingdom highlights this point in their document on public broadband schemes. "Ideally the competition should be at the infrastructure level, rather than solely at the service level, since this is likely to deliver the most choice and innovation for consumers".[32]

Despite this aim, most countries still rely on unbundling to ensure broadband competition. Even countries with both cable and DSL broadband networks such as Belgium have implemented unbundling and continue to rely on the increased competition it allows. In January 2007, Belgium's Telecommunications Consultation Committee reaffirmed that "unbundling the historic operator is necessary" in view of market competition, parti-cularly given how broadband has become "an important engine for the creation of substantial prosperity".[33]

In contrast, other countries with infrastructure competition have taken a more targeted approach to unbundling. For example, the United States still requires incumbent telephone companies to unbundle some types of legacy copper loops for competitive operators, but has capped unbundling in areas where the presence of unbundled network elements suggests that facilities-based competition is viable.[34] The United States has limited the obligation to unbundle access to next-generation networks based in part on the greater subscribership to cable modem service, as well as policy makers' concerns about possible investment disincentives.[35] Indeed, unbundling has been very successful in a number of OECD countries and less successful in others.

The majority of OECD member governments had implemented unbundling before the OECD Council Recommendation on Broadband Development. However, Switzerland and New Zealand both implemented unbundling of the copper loop during the previous two years in an effort to stimulate competition.

Unbundling debates are likely to become more pronounced as operators in countries with unbundled copper local loops upgrade their networks to fibre. Relatively few regulators have decided whether next-generation networks will be subject to unbundling requirements similar to those on copper local loops. The debates will be particularly important in countries without effective infrastructure-based competition.

Operational/functional/structural separation

Some policy makers have suggested that separating the ownership of infrastructure and the provision of services is one method for improving broadband competition in the economy. Supporters of structural separation argue that it would bring the incumbent's incentives into alignment with a non-integrated carrier. This could guarantee non-discriminatory access to (components of) the incumbent's networks and promote competitiveness which would then in turn promote innovation. Opponents of separation suggest that it would not be necessary or beneficial, that it would remove incentives to invest in new networks and that it would be costly to implement. Policy makers in some OECD countries have adopted functional separation while other governments have strongly opposed it.

In 2003, the OECD published a paper on structural separation. At the time, problems faced by new entrants in obtaining access to the network facilities of incumbents led to calls for structural remedies on incumbents, and in particular the separation of the local loop from service provision. The OECD report argued that the costs of structural separation would more than likely outweigh any benefits that it may provide.[36] However, while that conclusion may hold in the PSTN environment, it is not clear that it would still be valid in a fibre environment where high-entry costs may result in residents having access to only a single fibre network.

Since 2003, the increased interest in separation in several OECD countries has centred on "functional" separation (also called operational or accounting separation) of the retail and wholesale units of the incumbent. The policy objective is for the wholesale company to treat the incumbent's retail division the same way it treats all other wholesale customers. Regulators in several OECD countries have recently examined what role separation can play in improving this competitive climate.

Sweden's IT Policy Strategy Group poses the question of whether it is economically feasible to expect parallel infrastructure for high-speed broadband in Sweden. They find that parallel infrastructure works best in more densely populated areas but many rural and remote areas will likely have one (or no) high-speed infrastructure providers. In a number of areas of Sweden, neutral parties have installed fibre and lease it out to any service provider.[37] This distinction between the company putting in the infrastructure and those providing services forms the basis of discussions surrounding operational, functional, and structural separation. In 2007, Sweden's Post and Telecommunication Ministry issued its broadband strategy which included a proposed plan that PTS would require the functional (and/or) legal separation of TeliaSonera's wholesale unit.[38]

Australia introduced operational separation for the incumbent, Telstra, in 2003 as a way to ensure that Telstra's treatment of its own retail arm was equivalent to its treatment of wholesale customers. The separation requires that Telstra maintain separate wholesale and key network service business units.[39] BT in the United Kingdom also announced an operational separation in 2005. Ofcom promoted the separation as a way to allow all communications providers equal access to critical BT infrastructure on fair terms. Additionally, Ofcom announced that the separation would encourage investment in infrastructure, enable innovations through multiple services, and increase deployment of next-generation technology.[40]

Policy makers in New Zealand are also examining how to best implement operational separation. New Zealand's *Telecommunication Amendment Act* of December 2006 introduces operational separation for Telecom New Zealand and the Minister of Communications is currently developing a set of requirements.[41] The Turkish government has also examined what role structural separation could play in promoting broadband in the country.[42] The Information Society Action Plan 2006-2010 calls for a cost-benefit analysis to determine whether the separation of the wholesale and retail services of the incumbent operator would be justified.

The Czech Republic provides direct support to municipally owned networks which follow the "town owns the infrastructure" principle but where outside operators provide services.[43]

In certain cases the push for structural separation began with the owners rather than the regulator. In Ireland, the investment company which owns Eircom has expressed interest in selling off some or all of the company's retail assets.[44]

Regulators in other countries are looking for ways to avoid the costs of functional separation while still retaining similar outcomes. In July 2007, the French regulator ARCEP launched a public consultation on the sharing of indoor wiring in apartment buildings among operators.[45] One possible outcome is that operators will share the indoor wiring as a neutral access point and will instead compete on services. Similar arrangements have been successful in Korea where apartment buildings own their own internal wiring and allow multiple providers to connect in the basement to provide services.

Technologically neutral policy

Policy makers commonly use the phrases "non-discriminatory" or "technologically-neutral" in their documents to describe desired policies. This likely stems from a widely-held belief that governments should avoid helping or hindering certain technologies in the marketplace through regulation. As an example, Belgium's Telecommunication Consultative Committee recommends a balanced regulatory approach which is entirely neutral in terms of technology.[46] The OECD Council Recommendation on Broadband Development itself calls for "technologically neutral policy and regulation" as a way to "encourage interoperability, innovation and expand choice".

Policy makers in the United States, for example, have removed a number of technology-specific regulations in an effort to promote infrastructure investment by competing broadband platforms. This reflects the United States policy that providers of similar services must compete on a level playing field.

Technologically neutral regulation is a valid policy goal and market forces typically guide technology investment better than policy makers can. However, it is important to note that the competitive broadband markets in many OECD countries are actually the result of technologically-biased regulation. Unbundling requirements are applied to DSL lines in most OECD countries. They are not, however, applied to cable networks in OECD countries except for Canada. This brings into question whether technological neutrality is achievable – or even desirable – under current market conditions.

Where technological neutrality probably is more appropriate is in the financing of government projects. The Canadian expert group proposes a programme to finance broadband deployment. If implemented, the panel suggests:

"Because of the rapid evolution of technology, it is critical for U-CAN to be technologically neutral. The Panel believes there is great potential for the delivery of broadband to remote communities via new wireless access technologies such as WiMAX. However, no one can say for certain what technology will be the best two, three or five years from now. The Panel is recommending that the U-CAN program adopt a competitive technologically neutral approach. This approach should stimulate innovation and ensure that government subsidies are not used inefficiently or for obsolescent technologies."[47]

Other examples are the Spanish[48] and Hungarian governments who impose similar requirements on the dispersion of funds. The Hungarian Ministry of Economy and Transport position is:

"As to regulatory and deployment-funding measures related to broadband infrastructures and broadband access we must emphasise that a fundamental government principle in Hungary is technology-neutrality. This is true for both network solutions and end-user equipment. This approach is justified by the obligation to be unbiased in competition, the relative infancy of broadband solutions and the dynamism of technological development."[49]

There are important differences in the characteristics of various technologies used for Internet access and certain technologies may be more appropriate for different situations. Communities and governments should be able to maintain technological neutrality by requiring certain characteristics of the access connection, rather than a tender dictating a certain technology.

Consistent policies

The OECD Council Recommendation on Broadband Development highlights the need for consistent policies as a way to promote the convergence of platforms and services. Consistent policies create an atmosphere conducive to investment by the private sector, which is important for the development of telecommunication markets.

Government study groups formed by member countries have also emphasised the need for consistent policies as part of their overall broadband and ICT plans. The 2006 reports from the Swedish and Canadian study groups both express the importance of a consistent regulatory framework.

Sweden's IT Policy Strategy Group highlights the importance of predictability and a long-term approach. These apply to "regulations, interventions, ownership and the roles of the public sector".[50] In Canada, the expert group called for consistent policies, in particular across government agencies.

Consistent and clear policies are important for attracting investment and development in broadband markets in the OECD. The guidance from both Sweden and Canada suggests that policy makers should consider decisions carefully – with a long-term vision – rather than implementing shorter-term regulation to address specific market inefficiencies.

Promoting investment: Supply-side approaches

The Recommendation places a strong emphasis on promoting investment and infrastructure development. It states that Member countries should implement policies that encourage investment in new technological infrastructure. The Recommendation also suggests that member countries should implement demand and supply-based approaches in an effort to foster a virtual cycle of take-up and effective use of broadband. This section will examine a number ways governments have encouraged investment in new infrastructure.

Private and public-sector roles

Nearly all official broadband strategies across the OECD recognise and emphasise the role of the private sector in determining broadband market outcomes. As an example, policy makers in the Netherlands have designated the market with primary responsibility for investment in next generation broadband infrastructures. Policy actions in the Netherlands aim at "staying ahead in broadband infrastructure" and "boosting the use of broadband networks".[51]

The Japanese government has also promoted a private-sector-driven broadband market as part of their overall strategy. The strategy leaves the deployment of broadband services to private telecommunication carriers but the government helps promote the technologies using "appropriate policies and measures" that allow for "fair competition" and which maintain "investment incentives".[52]

Other countries have maintained that the private sector is the best to lead broadband efforts, however, that there are public and municipal entities which may also have the specialised skills necessary to build out networks and they may have the long-term financing strategies that are attractive for new rollouts.

In Norway, the government-supported Høykom programme has the mandate of inducing public sector institutions to acquire and use broadband connections and applications. The evaluation of the Høykom project found that many electricity companies were beginning to install telecommunications infrastructure themselves.

> *In response to the market situation, many local municipalities (whether they have Høykom project funding or not) have used their semi-privatised local electricity companies to build access networks for public and private customers alike. These companies have cabling skills, a complete local customer base, and a sound financial position. This has provided a quick solution for local authorities.*[53]

In general, policy makers in OECD countries have left the development of broadband infrastructure to the private sector. However, some governments and municipalities have quickly intervened in cases where private investment was taking too long to materialise or in areas where there was simply was no justified business case for a private firm but where social benefit was deemed high.

Reducing barriers to entry

As mentioned above, policy makers prefer that investment decisions be privately led. There are certain steps that governments have taken to improve the investment climate for infrastructure providers without interfering directly in markets. One of the key ways that policy makers have worked to achieve this is by reducing barriers to entry. Telecommunication and cable markets have long been considered high-entry barrier markets. Governments can *indirectly* help promote markets through reducing barriers to entry that are typically under their control or jurisdiction.

The broadband plan of the Czech Republic highlights the role that governments can play in reducing overall barriers to entry in broadband markets. The government takes a relatively "hands off" position with respect to government involvement with infrastructure investment but lists several areas where governments can help reduce barriers. One key element that the Czech Report indicates is "improving the legislative environment by removing various legislative and administrative obstacles which stand in the way of development competition in the electronic communications sector".[54]

The Canadian expert group also addresses the indirect nature of support by reducing entry barriers. However, they also warn that policy makers need to be cautious in the way that the government lowers entry barriers. They write, "Great care must be exercised in designing measures to lower barriers to entry, so as not to provide inappropriate incentives to both incumbents and new entrants".[55]

One way that local and national governments reduced barriers to entry was through funding backbone construction or providing access to backbone connectivity. The Recommendation does not specifically deal with backbone infrastructure but does encourage policies which "promote access on fair terms and at competitive prices to all communities, irrespective of location".

Sometimes policy makers have provided access to backhaul through regulatory means. The Australian regulator, the ACCC, identified backhaul routes where there is still limited competition. On these routes the ACCC regulated the transmission service, guaranteeing that other service providers can access the transmission routes at cost-based prices.[56]

Policy makers in other countries chose to fund high-speed backbone connectivity between institutions such as universities and research centres. In Iceland, the government's "Connect Policy" aims to connect all the principal government organisations with a high-speed, secure network. The project also aims to finance a high-capacity connection between Icelandic universities and research networks abroad.[57] The government in Luxembourg took a similar approach by connecting research institutions across the country using a minimum of 155 Mbit/s links and up to a theoretical limit of 1 Gbit/s. Luxembourg's policy also requires operators to install redundant capacity when they put in fibre networks.[58]

Korean policy makers have consistently emphasised backbone networks as part of their broadband strategy. Korea's initial investments to connect government offices and post offices across the country provided the necessary connectivity to reach rural areas and encouraged competitive access. This backbone infrastructure now serves as the foundation for Korea's broadband converged network (BcN).[59] Japan's policy makers link investments in backbone infrastructure to eliminating the digital divide and becoming "the world's most advanced ICT nation in 2010." The government goals lay out a plan to give more than 90% of households access to connections at 30 Mbit/s and higher. To this end, the government promoted the provision of optical fibres to regions without broadband connectivity as part of its IT New Reform Strategy.[60]

Policy makers in Greece also focussed on backbone networks in around 75 metropolitan areas. The government finances projects with the condition that they interconnect at least 20 spots of public interest in the given metropolitan area. In actuality, the metropolitan networks interconnect an average of 45 public-interest sites in each metropolitan network. Some bandwidth on the networks is set apart for private use via long-term leases with the aim that this will cover the networks operational and maintenance costs.[61]

As a way to promote broadband access and regional development, the Canadian government has been a strong supporter of backhaul connections to rural areas through programmes such as the National Satellite Initiate and the BRAND pilot programme. Various Provincial and Territorial governments in Canada have also implemented programmes for this purpose. In its work, Canada's expert panel found that providing backhaul from a local point of presence in a rural area back to national and international backbone networks was a much greater challenge than distributing connectivity to residences and businesses from the rural point of presence. As a result, the review panel suggests that bidding for government subsidies using its proposed U-CAN programme should be separate for backhaul and local access networks.[62]

Conduits, ducts, poles and rights of way

Operators have begun upgrading their copper networks to fibre, however, due to their high fixed costs, there are still significant concerns surrounding the economics of extensive new fibre installations. OECD research has found that civil costs (digging roads, installing on poles, etc.) can easily become the largest expense of fibre rollouts[63] (see Table 1.1). As a result, policy makers emphasise the important role that passive infrastructure (such as conduits, ducts and poles) plays in a number of country's broadband strategies.

For example, Italy's Broadband Committee proposes incentives that promote high-speed broadband by removing barriers to the use of civil infrastructure when operators lay fibre optic cabling.[64] In countries such as Canada there has been a move to eliminate rental fees from conduit usage in public areas or, as in the case of France, make them so inexpensive per metre that usage is essentially free. Belgian policy makers come to similar conclusions. They find that they can encourage investment by allowing access to conduit, sewers and roads and by reducing administration costs of the rollout.[65]

Table 1.1. Average estimated costs of installation of one kilometre of infrastructure, by type

Infrastructure type	Cost/km (water = 100%, 2002)
Water	100%
Sewer	35%
Electricity	26%
Gas	15%
Fibre optic	4-6%
Coaxial cable	2-4%
Copper wire	1-3%
Wireless connections	1-3%

Source: Upper Canada Networks via the publication "Broadband Electronic Communications in Hungary".[66]

Another way policy makers have encouraged investment and reduced costs is by co-ordinating infrastructure digs among providers. The costs per operator fall if operators are able to co-ordinate their builds in the same road. Sweden's IT Policy Strategy Group stresses that pipe-laying should be co-ordinated and that costs should be allocated among the providers installing lines.[67] The government can play an important role helping with the co-ordination among different market participants. However, policy makers must take care to not hold up the rollout of an operator for too long of a time solely to wait for other operators to join in the build.

Co-ordinating infrastructure digs allows operators to each lay down their own ducts and cables at the same time when the street is open. However, providers may be able to lower costs even further by sharing conduit or poles. Australian carriers are given special land access powers and some immunities from state and territory planning laws as a way to aid in infrastructure development and these same laws also encourage the sharing of passive infrastructure such as poles, ducts and towers.[68] Turkey's Information Society Strategy also encourages the installation of shared/common infrastructure by operators as way to promote broadband infrastructure development.[69]

Canada's expert panel explains the importance of improving access to and sharing passive infrastructure:

Wireline and wireless carriers require access to rights-of-way and support structures (e.g. poles, towers, and conduit). In addition, telecommunications service providers generally require access to in-building wiring in multi-unit buildings in order to supply services to customers. These elements are essential facilities. Without access to them, telecommunications service providers are unable to provision their networks or provide service to their end customers ... Furthermore, duplication of these facilities is uneconomic or undesirable. There has been increasing resistance from municipalities to the duplication of support structures. It is not in the public interest to have multiple sets of poles on streets or to have roads being dug up continually to accommodate multiple sets of underground ducts. It is also more economically efficient to share the costs of existing support structures than to duplicate this investment. Hence, these infrastructure elements are essential components of Canada's national telecommunications system.[70]

There are several other suggestions put forward to improve access to passive infrastructure. The Swedish IT Policy Group recommends that empty conduit is placed in the ground whenever excavation work is carried out.[71] Governments in other OECD countries implemented similar rules (see Box 1.2). The Swedish Policy Group suggests that the government should hold a review of the documentation rules to find a way to streamline the application and digging process.[72]

Box 1.2. Requiring developers to install a pit-and-pipe network in all new estates

To "avoid today's new land becoming tomorrow's broadband black spots", Australia's Whittlesea Council in Victoria has mandated the installation of an empty conduit (suitable for accommodating a future fibre network) across its area. The Whittlesea policy requires developers to install a pit-and-pipe network in all new estates. This network can be used in two ways: The developer can assume ownership provided that they take responsibility for securing the key outcomes sought by Whittlesea (advanced broadband services and infrastructure in a competitive framework).

Alternatively, carriers interested in offering their services to the community may lease access to the conduit from the Council on "friendly" terms.

Source: The State of Victoria's "Aurora Fibre to the Home Case Study".[73]

These types of interventions can have an impact on broadband development. A report by the United States Government Accountability Office found that access to rights-of-way, pole attachments and wireless-tower sites influenced the pace of infrastructure development, particularly at the local level.[74] This type of emphasis can be seen in the United States: in 2004, an Executive Memorandum from the President gave broadband providers more timely and cost-effective access to rights-of-way on Federal lands for their networks.[75]

Spectrum policy

While municipal rights of way can be a barrier for wired broadband, access to spectrum poses a similar barrier to entry for wireless firms wishing to provide broadband services. Wireless connectivity is increasingly important in OECD countries, particularly in rural and remote areas, and a number of national broadband plans highlight changes to spectrum policy which could have a positive effect on these services.

To take advantage of under-utilised spectrum in regional and remote areas the Australian government allocates apparatus licenses on an "over the counter" basis in the 1900-1920 and 2010-2025 MHz bands for use in regional and remote areas. The ACMA has issued more than 200 of these licenses since February 2005.[76] The use of these licenses allows wireless ISPs in rural areas to provide services quickly at low cost. Finland's broadband working group suggested the government reallocate the 450 MHz spectrum formerly used for Nordic Mobile Telephony (NMT) to wireless broadband. The 450 MHz frequency band offers mobility and long range connections so it is particularly suited to rural and remote usage. The Finnish government adopted the recommendation and awarded a national license in mid-2005.[77]

Policy makers in Australia, Canada, New Zealand, the United States and the United Kingdom are all moving towards spectrum trading and away from more "command and control" methods of spectrum management as a way to improve market efficiency.

The United Kingdom is one of a number of OECD countries in which policy makers are adapting their spectrum policies. Ofcom released its Spectrum Framework Review in June 2005 which provides for more open spectrum policies and promotes spectrum trading in secondary markets. Ofcom is very clear in the review, however, that broadband wireless access should not receive preferential treatment in spectrum allocations.

In line with its goals for light touch regulation and the use of market mechanisms, Ofcom does not believe that it is appropriate to regulate spectrum in such a manner as to favour BWA over other uses. Instead, Ofcom believes that it should make spectrum available for a range of uses such that BWA operators have as wide a choice as possible of the spectrum they might employ for their service. However, they will need to compete with other potential users of the spectrum. Some have noted that BWA might bring societal benefits and requested that Ofcom look favourably on spectrum provision as a result. Ofcom does not believe that in general it should promote societal benefits through spectrum policy, as discussed in Section 4.3. Such benefits are better delivered through intervening in the output market rather than the input market.[78]

While the Ofcom policy is technologically neutral, wireless broadband providers could still benefit from the ability to lease under-used spectrum in small geographic areas. Ofcom has stated that 72% of its spectrum under management should be available for market trading by 2010, 21% centrally managed and the remaining 7% should be licence-exempt.[79]

Clearly, the issue of access to spectrum for wireless broadband connections is going to be important in the coming years. There is no consensus yet on the best way to make new spectrum available but an increasing number of countries are looking into spectrum trading and secondary markets.

In some ways the amount of spectrum needed for wireless broadband is closely linked to the footprint of wired, high-capacity connections. New fibre connections can deliver high-speed access to more wireless antennae, which can then operate at lower power levels and ease some spectrum demand.

Other methods to remove barriers

There have been a number of other important measures taken in OECD countries to promote investment in infrastructure by reducing entry barriers.

* France eliminated individual licensing requirements and the requirement for local (municipal) authorisation for cable companies. Now providers can enter markets through a system of general authorisation. In addition, cable operators are no longer limited to a maximum size of 8 million households.

* The Hungarian Electronic Communications Act of 2004 reduced limitations on the number of subscribers allowed on certain networks. CATV service providers can now have up to one-third (instead of one-sixth) of

the total potential customer base. Then in 2007, Hungary abolished the provision.[80]

- Luxembourg modified its law (Law of 30 May 2005) to favour investment in communication networks as a way to stimulate competition.[81]

- Sweden's IT Policy Strategy Group suggested that the market should be encouraged to increase co-operation among the owners of physical communication network infrastructure and firms that supply electricity. They give particular emphasis to finding solutions for reserve power.[82] Ensuring an adequate power supply is necessary for network provision.

Government intervention

Determining whether government initiatives are appropriate and how they should be structured

One of the most important decisions facing policy makers is if and when to intervene in broadband markets. Broadband is beneficial to communities but there are questions about the role of public broadband schemes. A number of cities in the OECD have gone forward providing public Wi-Fi access and other municipalities are helping with the development of fibre connections to homes and businesses. National governments also subsidise backbone infrastructure to support broadband connections.

The regulator in the United Kingdom, Ofcom, emphasises that public broadband schemes have the potential to provide the benefits of broadband to rural and remote areas. Many of these areas would otherwise be disadvantaged without the connectivity that makes businesses competitive and gives end-users access to Internet services. Ofcom makes a point, however, that "it is critical that these schemes are well targeted and well structured, or they risk distorting competition between commercial operators, and/or resulting in avoidable use of public funds.[83]

Public entities across the OECD intervene in very different ways in broadband markets. Some municipalities build their own networks while other governments take a very market-based approach. Decision makers should examine each new proposal for government intervention individually, and there are some key guidelines which local and national governments should follow.

There seems to be a near universal acceptance in OECD markets that reducing administrative barriers to entry in the broadband network market is a relatively good way to promote investment. This can be done by removing various legislative and administrative obstacles that stand of the way of developing competitive networks in the communication sector. In a more general way, any improvements to the investment climate overall in the economy will help entice network infrastructure development. The Czech Republic's broadband plan highlights both of these key points.[84]

Governments also need to be very aware of the areas where broadband competition is sufficient and other areas which lack sufficient coverage as a result of market forces. Italy's broadband committee[85] and Sweden's IT Policy Strategy Group[86] have devised detailed coverage maps showing where broadband is available and areas where it is not. Norway has also published interactive coverage maps.[87] These maps could serve as a guide for operators and other actors wishing to establish themselves in particular areas. They would also give policy makers an idea of where public intervention could have the most impact with the least market distortion or where intervention may not be appropriate. In their IT Strategy, Japanese policy makers are very explicit that private-sector incentives can be used to increase broadband availability as long as they can still maintain fair competition. Japanese policy makers also look favourably upon the development and sharing of regional public networks.[88]

Once policy makers decide there are grounds for intervention there are various ways to ensure that projects are as efficiently chosen as possible. Canada's expert panel suggests that "least-cost subsidy auctions" can be used to select the most efficient and qualified service providers to build the network to a given set of criteria. They recommend that the bidding process be technologically neutral so that operators are allowed to propose the most "efficient and effective technologies available" to meet the requirements of the project.[89]

Some governments put restrictions on the bidding process for public communications networks in an effort to ensure that the subsidised networks are for the public good (see Box 1.3). Policy makers in the Czech Republic have set specific criteria for groups looking to use structural funds to build communications networks. Only public bodies such as municipalities, regions (or organisations and associations created by regions and municipalities) or non-governmental non-profit organisations are allowed to apply for funds. In addition, the recipients of funding must own everything that is constructed using the money; they must also remain owners of the network for a minimum of five years.[90]

Even if policy makers decide that intervention is justified, government policy makers should continuously monitor broadband markets to ensure that government participation is still necessary. Sweden is home to many municipally-run broadband networks and the regulator, PTS, has taken steps to ensure that these networks don't distort the market. If municipalities own networks where commercial firms have installed "future-proof" broadband infrastructure, or even where they might do so, PTS recommends that the municipality dispose of the investment or take steps to ensure that competition is not distorted.[91]

Box 1.3. Rules for getting funding in Hungary

The New Hungary Development Plan lays out specific criteria for firms applying for structural funds to expand broadband coverage using Hungarian national sources and structural funds of the European Union.

These include:

- Minimum 2048/512 kbit/s (download/upload speeds) but a symmetric 2 Mbit/s is preferred.

- Service availability must be at least 97% according to the government decree 345/2004.

- No time limitations for connections.

- The capacity of the network must be adjustable to accommodate subscriber's bandwidth demand.

- In the maintenance period, all subscriber demand must be settled within 60 days.

- Tenders which provide for *(1)* installation of local public terminals and equipment and *(2)* organisation of training for the local population and enterprises to encourage ICT and broadband Internet use are preferred.

Source: Hungarian Ministry of Economy and Transport.[92]

Open access

When governments do decide to intervene in markets by subsidising communication networks, there is often a clause that stipulates that the network must be "open access". The term "open access" refers to an arrangement where network providers offer capacity or access to all market participants under the same terms and conditions. Operators of open access networks must allow competitive access to the network on non-discriminatory terms.

An increasing number of governments and policy review groups are calling for open access requirements as a condition of funding.

* All Canadian telecommunications infrastructure projects that received funding from the BRAND Pilot Program were required to be open to third-party providers of local access services, as a condition of receiving such funding. The Canadian expert group recommended that the same open access conditions be applied to all new backhaul networks which receive funding under their proposed U-CAN Program. The Panel went a step further by suggesting that the rates charged to third parties for access "should be discounted to reflect the subsidies received and to ensure a level playing field between competing service providers".[93]

* The Spanish Ministry of Industry, Tourism and Commerce requires "open access to the network for other operators" as a condition for receiving funding for broadband rollouts.[94]

* Sweden's regulator, PTS, put forward three possible trajectories for the Swedish broadband plan, one of which emphasises that any broadband networks financed with "central government support" should be open to service providers other than the network owner during the (entire) lifetime of the networks.

Some governments have made the development of open access fibre networks a key element of their overall broadband strategies. One of the goals of New Zealand's Digital Strategy Programme and Broadband Challenge is to encourage the development of open-access optical fibre networks in urban areas and the provision of affordable broadband services in rural and underserved areas.[95]

Policy makers have also extended open access requirements beyond projects which receive government funding. In cases of substantial economic power in markets, policy makers have applied open access rules to existing carriers. The Cabinet of Ministers in the Netherlands has introduced a new *Telecommunications Act* which regulates open access and states that access obligations may be imposed on providers if they have an economic power position that gives rise to "undesirable situations".[96]

Complementary government initiatives to expand coverage

As highlighted earlier, there are still large areas in the OECD that do not yet have broadband access. The percentage of the population with access to broadband may be increasing but rural and remote communities still have limited or no options available to them.

The Broadband Stakeholder Group in the United Kingdom makes an argument that there are situations where the public value of next-generation networks (broadband) for society and the economy as a whole is potentially high but the potential value to private investors can be relatively weak. This is because of a combination of the large scale of investments and the uncertainties surrounding the prospects for recouping the investment.[97] This would indicate there is a possible role for government intervention.

At the same time, policy makers must take special care not to disrupt the normal functioning of markets when they decide to intervene and promote investment. There are no precise rules for when government funding would be acceptable and when it should be avoided. However, it is clear that there are circumstances when governments should help spur investment in broadband through various financial measures. Previous OECD research cautions governments against using a "blunt blanket" approach of adding broadband connectivity to existing universal service obligations[98] instead of a more targeted approach to covering broadband "black holes". Many regulators in OECD member countries have come to the conclusion that high-speed broadband access should not, at present, be a component of universal service obligations.[99]

That is not to say that universal service funds have not helped bring broadband connections to certain areas. In the United States, for example, universal service funds have helped increase broadband penetration by providing funding for new lines in rural areas.[100] Communities benefit as long as the local loop is short enough to support broadband. There are examples of countries where broadband service has been made a part of universal service obligations. The Swiss government has decided to include broadband in universal service as a way to expand coverage. From 1 January 2008, the entire Swiss population will be able to have broadband access. The requirements state that connections must offer at least 600 kbit/s downloads and 100 kbit/s uploads and the monthly subscription cannot be more than CHF 69.[101]

Policy makers do not commonly address the need for broadband infrastructure through universal service requirements. Instead, they use a number of methods to target specific areas in the greatest need of broadband connections. Support is commonly provided through private/public sector partnerships, interest-free loans or tax concessions.

There are a large number of current projects in the OECD that range from national broadband subsidies to small community networks in villages. In order to narrow the discussion, this section will examine just a sample of national interventions designed to improve broadband infrastructure development. The section will finish by summarising overall trends and findings of the section.

1. New Zealand – "Broadband Challenge". The goal of the programme is encouraging the development of open-access optical fibre networks in urban and underserved areas. A total of NZD 17.9 million was awarded to successful projects. The projects typically have a high level of community support as local institutions (District Health Boards, tertiary institutions, schools and local businesses) are "buying in" to the project.[102]

2. Norway – "Broadband for All". The Norwegian government's objective with the plan is to offer all citizens of Norway a connection to broadband by the end of 2007. The Høykom Programme is one of the key instruments used to achieve the goal. The programme lays out very precise criteria for funding and reaches out to municipalities, local businesses and firms in Norway's regions. A new sub-programme called "Høykom district" promotes broadband for outlying districts. In 2007, the Norwegian government had spent a total of NOK 355 million to subsidise the establishment of broadband infrastructure in areas with no existing broadband offer. These State funds must be combined with at least 50% local funding and it has been estimated that the broadband coverage in Norway will reach 99% (fixed access) when these combined funds are invested. In addition, there are mobile broadband solutions with increasing speed and coverage that – if included – extend the reach beyond 99% of households.[103]

3. Korea – "Broadband Convergence Network". Policy makers and telecommunication firms in Korea are pushing forward with the development of a converged broadband network called the BcN. The Ministry of Information and Communication generated private investment worth KRW 12.8 trillion for the first phase (2004-2005). In 2006, the Informatization Promotion Committee established the "Basic Plan for Establishing Broadband Convergence Network II" and planned for early construction of BcN infrastructure with the goal of promoting a ubiquitous network.[104]

4. Italy – "Financial Act". Italy's Financial Act assigns an additional EUR 10 million a year for 2007, 2008 and 2009 to help develop broadband in Southern Italy. The project relies on the Ministry of Communications and Infratel Italia.

5. United States – "Rural Utilities Service". The United States Department of Agriculture has a programme called the Rural Utilities Service (RUS) which provides grants to improve rural infrastructure. Some programmes within this service are aimed at providing broadband service. The Community Connect Program

offers grants to deploy transmission infrastructures for broadband service in communities where no broadband services exist. The programme requires grantees to wire specific community facilities and provide free access to broadband services in those facilities for at least two years. Grants are targeted at low population, rural areas and can be awarded to entities that want to serve a rural area of fewer than 20 000 residents. In 2007, the RUS awarded USD 10.3 million in Community Connect grants. Various state and local-level programmes also exist.

6. Ireland – "National Broadband Scheme". Ireland's National Broadband Scheme (NBS) is a government project funded under the National Development Plan to provide nation-wide broadband coverage. Under the scheme, connections to users need to be an "always-on" connection capable of 1 Mbit/s downloads and 128 kbit/s uploads. The lowest possible bit cap on connections will be 10 giga-bytes per month and the connections must support virtual private networks and VoIP applications. The project is technologically neutral and the service provider will be engaged for a period of five years.[105]

7. Spain – Public/Private partnerships. The Spanish government put together a public/private partnership with the goal of improving broadband connectivity in Spain. Of the public funds contributed, EUR 31 million were structural funds and EUR 53 million were in zero-interest public credits. Operators themselves invested an estimated EUR 280 million. The funded projects use ADSL, WiMAX and satellite technologies depending on geography, roll-out dates and the available technologies. The government set the minimum download speed at 256 kbit/s. Prices were also capped at a "reasonable fee". The project connected 3 700 communities in the first phase. As a related intervention, the government is also offering zero-interest credits to buy computer equipment and broadband connections for businesses and consumers. These credits amount to EUR 700 million in 2007.[106]

8. Australia –"Connect Australia". In 2005, the Australian government announced an AUD 1.1 billion communications package for regional access to telecommunications services. The government also established an AUD 2 billion Communications Fund to finance government responses to future legislative reviews of regional telecommunications.[107] In June 2007, the former Australian government announced that OPEL Networks won a contract to provide broadband connectivity across the country using a combination of DSL and WiMAX technologies. The new network is planned to

cover 638 000 square kilometres (extending across all States and Territories) and will be completed in 2009. Broadband speeds will initially be up to 6 Mbit/s, rising to 12 Mbit/s using wireless connections and 20 Mbit/s using DSL by 2009.[108]

9. Czech Republic – "Broadband subsidy title". The Czech government put 1% of the proceeds from the privatisation of Český Telecom into a fund that will be used to co-finance infrastructure projects for metropolitan and local networks. Conditions on receiving the funds include participation by the relevant regions and that the network operated under "open access" rules. In 2006 the government awarded over EUR 5 million to 47 different projects.[109]

10. Hungary – Tax concessions. Hungary's government grants a tax reduction of 50% on profits as a way to support the construction of broadband infrastructure. The concessions are available only to tele-communication companies if their expected profits exceed HUF 50 million and if they have invested at least HUF 100 million. The tax allowance cannot be applied to ISPs if the infrastructure is built in areas where Internet service is already provided or where the investment does not contribute to the growth in infrastructure. The government has also taken steps to subsidise end-user equipment for educators, students and low-income families. Under the "Sulinet Expressz" project, over 500 000 products were purchased, 60 000 new computers were installed in homes and 150 000 computers were upgraded.[110] This latter tax concession was abolished in 2006.

11. Turkey – Public Internet access centres. The Turkish Information Society Strategy[111] and its annexed Action Plan[112] endeavour to establish public Internet access centres across Turkey to provide computer and Internet access to those who do not have access at home. The strategy targets libraries, public foundations, corporations, municipalities, organised industrial regions, public training centres and volunteered foundation buildings as potential locations to provide access to citizens. A public-private partnership with Turk Telecom established 716 public Internet access centres in various districts. The government is currently looking into expanding the number of public Internet access centres across the country. In addition, the government has extended connectivity to military conscripts though a total of 227 Public Internet Access Centres with 4 487 computers, 227 projectors, printers and related equipment in military campuses.

Over the previous 3-4 years, there have been a large number of government interventions to expand coverage in broadband markets. Interventions at the national level have tended to support the rollout of backbone infrastructure (discussed earlier in the paper under the barriers to entry section) or have been public/private schemes to rollout broadband in certain geographic areas.

Some public infrastructure rollouts have been very successful, particularly a number of fibre-to-the-home rollouts in small communities in the Netherlands. Other public interventions have been more problematic – such as city-wide Wi-Fi initiatives in a number of OECD countries.

One of the difficulties with government interventions is that policy makers must decide on the characteristics of a "basic connection" for the project at a time when advances in broadband technologies, speeds and applications make defining a "basic connection" very difficult. In this sense, policy makers may wish to look for technologies which offer more "future-proof" solutions which can be easily upgraded as markets advance.

Notes

[1]   Greece is an example of a country which has used penetration rates to benchmark its broadband progress. "The Plan for the Development of Broadband Services until 2008", Ministry of Economics and Finance – Greece, at: www.infosoc.gr/infosoc/en-UK/specialreports/broadband_plan/default.htm.

[2]   Comments submitted to the OECD by Luxembourg, July 2007.

[3]   "CRTC Telecommunications Monitoring Report", Canadian Radio-television and Telecommunications Commission, July 2007, at: www.crtc.gc.ca/eng/publications/reports/PolicyMonitoring/2007/tmr2007.pdf.

[4]   "High-Speed Services for Internet Access: Status as of December 31, 2006," Federal Communications Commission – United States, October 2007 at: http://hraunfoss.fcc.gov/edocs_public/attachmatch/DOC-277784A1.pdf.

[5]   "Broadband Coverage in Europe - Final Report - 2006 Survey", IDATE, November 2006, at: http://ec.europa.eu/information_society/eeurope/i2010/docs/benchmarking/broadband_coverage_06_2006.doc.

[6]   OECD (2007) *OECD Communications Outlook 2007*. OECD Publishing, Paris.

[7]   "The Fibre Battle", JP Morgan European Equity Research, 04 December 2006.

[8]   "The Netherlands: FTTH deployment overview 4Q2006", Stratix, Jan 2007 at: www.stratix.nl/documents/FTTH-B-C_overview_final.pdf.

[9]   "Vienna plans FTTH to 960,000 households",CAnet, 26 January 2006 at: http://lists.canarie.ca/pipermail/news/2006/000203.html.

[10]   "Understanding Broadband Deployment in Vermont", Vermont Department of Public Service, February 2007 at: http://publicservice.vermont.gov/Broadband/Broadband%20Deployment%20in%20Vermont%20Final.pdf.

[11]   OECD (2007) *OECD Communications Outlook 2007*. OECD Publishing, Paris.

[12]   "Broadband Connectivity and the Role of Satellite Solutions to Address the Digital Divide", European Satellite Operators Association submission to the OECD via BIAC, June 2007.

[13]   "About us - Broadband for Rural and Northern Development Pilot Program", Industry Canada, at: http://broadband.gc.ca/pub/program/about.html.

[14]   "South Korea launches WiBro service", EE Times, 30 June 2006 at: www.eetimes.com/news/latest/showArticle.jhtml?articleID=189800030.

[15]   OECD (2007) *OECD Communications Outlook 2007*. OECD Publishing, Paris.

[16]   OECD (2005-2007) *OECD Communications Outlook 2005-2007*. OECD Publishing, Paris.

[17]   *Ibid.*

[18] "The Commission's Broadband for all" policy to foster growth and jobs in Europe: Frequently Asked Questions, European Commission Press Release, 21 March 2006 at:
http://europa.eu/rapid/pressReleasesAction.do?reference=MEMO/06/132&format=HTML&aged=1&language=EN&guiLanguage=en

[19] *Ibid.*

[20] "HSPA: High Speed Wireless Broadband - From HSDPA to HSUPA and Beyond", UMTS Forum at: www.umts-forum.org/servlet/dycon/ztumts/umts/Live/en/umts/MultiMedia_PDFs_Papers_White-Paper-HSPA.pdf.

[21] "Wireless Last Mile Final Report", Ofcom – United Kingdom, 20 Nov 2006, at: www.ofcom.org.uk/research/technology/overview/ese/lastmile/lastmile1.pdf.

[22] "The National Broadband Access Policy", Czech Republic Ministry of Informatics, 2005, at: www.micr.cz/files/2185/MICR_brozura_en.pdf.

[23] "Wireless Last Mile Final Report", Ofcom – United Kingdom, 20 Nov 2006, at: www.ofcom.org.uk/research/technology/overview/ese/lastmile/lastmile1.pdf.

[24] "Internet Service Provider Survey", Statistics New Zealand, 7 March 2007, at: www.stats.govt.nz/NR/rdonlyres/D9A1935F-E437-4DE4-8E0B-1ACD601700C3/0/internetserviceprovidersurveysep06hotp.pdf.

[25] "The National Broadband Access Policy", Czech Republic Ministry of Informatics, 2005, at: www.micr.cz/files/2185/MICR_brozura_en.pdf.

[26] "What other communities are doing", Ottawa 2020 Broadband Plan, at: http://ottawa.ca/city_services/planningzoning/2020/bb/appen_d_en.shtml.

[27] The term unbundling in telecommunications refers to a regulatory requirement that the historical telecommunication operator make its copper lines available to competitors for a set monthly fee so they can provide phone and Internet services.

[28] Effective competition is commonly defined by economists as the persistent absence of players with market power, where market power is the ability to influence prices and persistently realise higher profits than those firms which do not possess market power. Users are better off in an effectively competitive market because they are more likely to be provided a greater variety of products and/or services of lower price and higher quality than they could be provided in a non-competitive market.

[29] "Telecommunications Policy Review Panel Final Report 2006", Industry Canada, March 2006, at:
www.telecomreview.ca/epic/site/tprp-gecrt.nsf/vwapj/00A_e.pdf/$FILE/00A_e.pdf.

[30] "FCC Electronic Comment Filing System", Federal Communication Commission – United States, at: www.fcc.gov/cgb/ecfs/.

[31] "Telecommunications Policy Review Panel Final Report 2006", Industry Canada, March 2006, at:
www.telecomreview.ca/epic/site/tprp-gecrt.nsf/vwapj/00A_e.pdf/$FILE/00A_e.pdf.

[32] "Public Broadband Schemes: A Best Practice Guide", OFCOM – United Kingdom, February 2007, at: www.berr.gov.uk/files/file37744.pdf.

[33] Opinion of the Consultative Committee on Telecommunications of 31/01/2007 "How to improve broadband penetration in Belgium?", at: Dutch version: www.rct-cct.be/docs/adviezen/NL/070131%20breedband%20nl.pdf; French version:

www.rct-cct.be/docs/adviezen/FR/070131%20breedband%20fr.pdf.

[34] United States Court of Appeals for the District of Columbia Circuit, No. 05-1095, Covad Communications Company and Dieca Communication vs. Federal Communications Commission and United States of America, at: http://hraunfoss.fcc.gov/edocs_public/attachmatch/DOC-266208A1.doc.

[35] "Review of the Section 251 Unbundling Obligations of Incumbent Local Exchange Carriers; Implement of the Local Competition Provisions of the Telecommunications Act of 1996; Deployment of Wireline Services Offering Advanced Telecommunications Capability; Appropriate Framework for Broadband Access to the Internet over Wireline Facilities, Report and Order and Order on Remand and Further Notice of Proposed Rulemaking", FCC 03-36, Paras. 247-97 (rel. Aug. 21, 2003) available at http://hraunfoss.fcc.gov/edocs_public/attachmatch/FCC-03-36A1.pdf.

[36] "The Benefits and Costs of Structural Separation of the Local Loop", OECD DSTI/ICCP/TISP(2002)13/FINAL, 03 November 2007 at: www.oecd.org/dataoecd/39/63/18518340.pdf.

[37] "Broadband for growth, innovation and competitiveness" The IT Policy Strategy Group – Sweden, at: www.sweden.gov.se/sb/d/574/a/76048;jsessionid=aRgt9J6DAf-g.

[38] "Förslag till bredbandsstrategi för Sverige", Post- och telestyrelsen, 15 February 2007, at: www.pts.se/Dokument/dokument.asp?ItemId=6541.

[39] "Broadband Blueprint—Broadband Development", Department of Communications, Information Technology and the Arts – Australia, 2006: at: www.dcita.gov.au/communications_for_consumers/internet/broadband_blueprint/broadband_blueprint.

[40] "Ofcom accepts undertakings from Board of BT Group plc on operational separation", OFCOM – United Kingdom, 22 September 2005 at: www.ofcom.org.uk/media/news/2005/09/nr_20050922.

[41] "Operational Separation of Telecom", Ministry of Economic Development – New Zealand, Accessed 21 August 2007, at: www.med.govt.nz/templates/ContentTopicSummary____26310.aspx.

[42] "Information Society Action Plan 2006-2010", State Planning Organisation – Turkey, July 2006 at: www.dpt.gov.tr/konj/DPT_Tanitim/pdf/Information_Society_Strategy.pdf.

[43] "The National Broadband Access Policy", Czech Republic Ministry of Informatics, 2005, at: www.micr.cz/files/2185/MICR_brozura_en.pdf.

[44] "Eircom break-up backed by FF", The Post – Ireland, 06 May 2007, at: http://archives.tcm.ie/businesspost/2007/05/06/story23448.asp.

[45] "Mutualisation de la partie terminale des réseaux de fibre", ARCEP, 26 July 2007 at: www.arcep.fr/index.php?id=8571&tx_gsactualite_pi1[uid]=964&tx_gsactualite_pi1[annee]=&tx_gsactualite_pi1[theme]=&tx_gsactualite_pi1[motscle]=&tx_gsactualite_pi1[backID]=26&cHash=69dda645c6.

[46] Opinion of the Consultative Committee on Telecommunications of 31/01/2007 "How to improve broadband penetration in Belgium?", at: Dutch version: www.rct-cct.be/docs/adviezen/NL/070131%20breedband%20nl.pdf; French version: www.rct-cct.be/docs/adviezen/FR/070131%20breedband%20fr.pdf.

47   "Telecommunications Policy Review Panel Final Report 2006", Industry Canada,
       March 2006, at:
       www.telecomreview.ca/epic/site/tprp-
       gecrt.nsf/vwapj/00A_e.pdf/$FILE/00A_e.pdf.

48   "Bridging the Broadband Gap: Spanish Ministry of Industry, Tourism and
       Commerce, 15th May 2007, at:
       www.bandaancha.es/EstrategiaBandaAncha/ProgramaExtensionBandaAnchaZona
       sRuralesAisladas/EnglishInformation/.

49   "Broadband Electronic Communications in Hungary", Ministry of Informatics and
       Communications – Hungary, at:
       www.itktb.hu/resource.aspx?ResourceID=Broadband_Electronic_Communication
       s_in_Hungary_V1.

50   "Policy for the IT society: Recommendations from the members of the IT Policy
       Strategy Group", IT Policy Strategy Group – Sweden, 26 October 2006, at:
       www.sweden.gov.se/sb/d/574/a/76046;jsessionid=aRgt9J6DAf-g/.

51   "Broadband and Grids Technology in the Netherlands", Ministry of Economic
       Affairs, 2005 at: www.hightechconnections.org/2005/broadband.pdf.

52   "The Next-Generation Broadband Strategy 2010", Ministry of Internal Affairs
       and Communications - Japan, August 2006.

53   "HØYKOM Introduction ", Available at:
       www.hoykom.no/hoykom/hoykomweb.nsf/4a87ff3bf2c03cc38525646f0072ffa9/4
       2de84fc9e2014aac125700c00497969/$FILE/Evaluation%20_e.pdf.

54   "Telecommunications Policy Review Panel Final Report 2006", Industry Canada,
       March 2006, at:
       www.telecomreview.ca/epic/site/tprp-
       gecrt.nsf/vwapj/00A_e.pdf/$FILE/00A_e.pdf.

55   "Telecommunications Policy Review Panel Final Report 2006", Industry Canada,
       March 2006, at:
       www.telecomreview.ca/epic/site/tprp-
       gecrt.nsf/vwapj/00A_e.pdf/$FILE/00A_e.pdf.

56   "Broadband Blueprint—Broadband Development", Department of
       Communications, Information Technology and the Arts – Australia, 2006 at:
       www.dcita.gov.au/communications_for_consumers/internet/broadband_blueprint/
       broadband_blueprint.

57   "Resources to Serve Everyone: Policy of the Government of Iceland on the
       Information Society", Prime Minister's Office – Iceland, 2004, at:
       http://eng.forsaetisraduneyti.is/media/English/IT_Policy2004.pdf.

58   "La pénétration des technologies de l'information au Luxembourg", Mindforest,
       at: www.eco.public.lu/documentation/publications/reperes/Reperes.pdf.

59   "2006 Korea Internet White Paper", Ministry of Information and Communication
       – Korea, May 2006, at:
       http://eng.mic.go.kr/eng/secureDN.tdf?seq=10&idx=1&board_id=E_04_03.

60   "White Paper 2006 - Information and Communications in Japan", Ministry of
       Internal Affairs and Communications – Japan, at:
       www.johotsusintokei.soumu.go.jp/whitepaper/eng/WP2006/chapter-3.pdf.

61   "The Plan for the Development of Broadband Services until 2008", Ministry of
       Economics and Finance – Greece, at:
       www.infosoc.gr/infosoc/en-UK/specialreports/broadband_plan/default.htm.

[62] "Telecommunications Policy Review Panel Final Report 2006", Industry Canada, March 2006, at: www.telecomreview.ca/epic/site/tprp-gecrt.nsf/vwapj/00A_e.pdf/$FILE/00A_e.pdf.

[63] "Developments in Fibre Technologies and Investment", OECD DSTI/ICCP/CISP(2007)4/REV1.

[64] "Il Comitato Banda Larga", Ministero delle Comunicazioni - Ministero degli Affari Regionali e Autonomie Locali Ministero delle Riforme e Innovazioni nella Pubblica Amministrazione, at: www.comitatobandalarga.it/news/49/10/il_comitato_banda_larga.html.

[65] Opinion of the Consultative Committee on Telecommunications of 31/01/2007 "How to improve broadband penetration in Belgium?", at: Dutch version: www.rct-cct.be/docs/adviezen/NL/070131%20breedband%20nl.pdf; French version: www.rct-cct.be/docs/adviezen/FR/070131%20breedband%20fr.pdf.

[66] "Broadband Electronic Communications in Hungary", Ministry of Informatics and Communications – Hungary, at: www.itktb.hu/resource.aspx?ResourceID=Broadband_Electronic_Communications_in_Hungary_V1.

[67] "Broadband for growth, innovation and competitiveness" The IT Policy Strategy Group – Sweden, at: www.sweden.gov.se/sb/d/574/a/76048;jsessionid=aRgt9J6DAf-g.

[68] "Broadband Blueprint—Broadband Development", Department of Communications, Information Technology and the Arts – Australia, 2006: at: www.dcita.gov.au/communications_for_consumers/internet/broadband_blueprint/broadband_blueprint.

[69] "Information Society Strategy 2006-2010", State Planning Organisation – Turkey, July 2006 at: www.dpt.gov.tr/konj/DPT_Tanitim/pdf/Information_Society_Strategy.pdf.

[70] "Telecommunications Policy Review Panel Final Report 2006", Industry Canada, March 2006, at: www.telecomreview.ca/epic/site/tprp-gecrt.nsf/vwapj/00A_e.pdf/$FILE/00A_e.pdf.

[71] "Broadband for growth, innovation and competitiveness" The IT Policy Strategy Group – Sweden, at: www.sweden.gov.se/sb/d/574/a/76048;jsessionid=aRgt9J6DAf-g.

[72] "Policy for the IT society: Recommendations from the members of the IT Policy Strategy Group", IT Policy Strategy Group – Sweden, 26 October 2006, at: www.sweden.gov.se/sb/d/574/a/76046;jsessionid=aRgt9J6DAf-g/.

[73] "Aurora Fibre-to-the-Home Case Study", State of Victoria – Australia, August 2006 at: www.mmv.vic.gov.au/uploads/downloads/BAO/aurora110906.pdf.

[74] "Broadband Deployment Is Extensive throughout the United States, but It Is Difficult to Assess the Extent of Deployment Gaps in Rural Areas", United States Government Accountability Office, 5 May 2006 at: www.gao.gov/new.items/d06426.pdf

[75] See: www.whitehouse.gov/news/releases/2004/04/20040426-2.html. See also Federal Working Group report at: www.ntia.doc.gov/reports/fedrow/index.html.

[76] "Broadband Blueprint—Broadband Development", Department of Communications, Information Technology and the Arts – Australia, 2006: at:

www.dcita.gov.au/communications_for_consumers/internet/broadband_blueprint/
broadband_blueprint.

77 "National broadband strategy: Final report", Ministry of Transport and
Communications Finland, 23 January 2007 at: www.mintc.fi/oliver/upl615-
LVM11_2007.pdf.

78 "Spectrum Framework Review", Ofcom – United Kingdom, 28 June 2005, at:
www.ofcom.org.uk/consult/condocs/sfr/sfr/sfr_statement.

79 "Spectrum Framework Review", Ofcom – United Kingdom, 28 June 2005, at:
www.ofcom.org.uk/consult/condocs/sfr/sfr/sfr_statement.

80 "Broadband Electronic Communications in Hungary", Ministry of Informatics and
Communications – Hungary, at:
www.itktb.hu/resource.aspx?ResourceID=Broadband_Electronic_Communication
s_in_Hungary_V1.

81 "La pénétration des technologies de l'information au Luxembourg", Mindforest,
at: www.eco.public.lu/documentation/publications/reperes/Reperes.pdf.

82 "Policy for the IT society: Recommendations from the members of the IT Policy
Strategy Group", IT Policy Strategy Group – Sweden, 26 October 2006, at:
www.sweden.gov.se/sb/d/574/a/76046;jsessionid=aRgt9J6DAf-g/.

83 "Public Broadband Schemes: A Best Practice Guide", OFCOM – United
Kingdom, February 2007, at: www.berr.gov.uk/files/file37744.pdf"Pipe Dreams?
Prospects for next generation broadband deployment in the UK", Broadband
Stakeholder Group, 16 April 2007, at:
www.broadbanduk.org/component/option,com_docman/task,doc_download/Itemi
d,7/gid,930/.

84 "The National Broadband Access Policy", Czech Republic Ministry of
Informatics, 2005, at: www.micr.cz/files/2185/MICR_brozura_en.pdf.

85 "Il Comitato Banda Larga", Ministero delle Comunicazioni - Ministero degli
Affari Regionali e Autonomie Locali Ministero delle Riforme e Innovazioni nella
Pubblica Amministrazione, at:
www.comitatobandalarga.it/news/49/10/il_comitato_banda_larga.html.

86 "Broadband for growth, innovation and competitiveness" The IT Policy Strategy
Group – Sweden, at:
www.sweden.gov.se/sb/d/574/a/76048;jsessionid=aRgt9J6DAf-g.

87 "Dekningskart bredbånd", at: http://85.19.158.171/Mariaweb/HKOM/.

88 "New IT Reform Strategy: Realizing Ubiquitous and Universal Network Society
Where Everyone Can Enjoy the Benefits of IT", IT Strategic Headquarters –
Japan , 19 January 2006, at:
www.kantei.go.jp/foreign/policy/it/ITstrategy2006.pdf.

89 "Telecommunications Policy Review Panel Final Report 2006", Industry Canada,
March 2006, at: www.telecomreview.ca/epic/site/tprp-
gecrt.nsf/vwapj/00A_e.pdf/$FILE/00A_e.pdf.

90 "The National Broadband Access Policy", Czech Republic Ministry of
Informatics, 2005, at: www.micr.cz/files/2185/MICR_brozura_en.pdf.

91 "Förslag till bredbandsstrategi för Sverige", Post- och telestyrelsen, 15 February
2007, at: www.pts.se/Dokument/dokument.asp?ItemId=6541.

92 "Broadband Electronic Communications in Hungary", Ministry of Informatics and
Communications – Hungary, at:
www.itktb.hu/resource.aspx?ResourceID=Broadband_Electronic_Communication
s_in_Hungary_V1.

[93] "Telecommunications Policy Review Panel Final Report 2006", Industry Canada, March 2006, at: www.telecomreview.ca/epic/site/tprp-gecrt.nsf/vwapj/00A_e.pdf/$FILE/00A_e.pdf.

[94] "Bridging the Broadband Gap: Spanish Ministry of Industry, Tourism and Commerce, 15th May 2007, at: www.bandaancha.es/EstrategiaBandaAncha/ProgramaExtensionBandaAnchaZonasRuralesAisladas/EnglishInformation/.

[95] "Broadband Challenge Update Dec 06: What is happening now?", New Zealand Digital Strategy, accessed on 21 August 2007 at: www.digitalstrategy.govt.nz/templates/Page____928.aspx.

[96] "Better Performance with ICT: Update of the ICT Agenda of the Netherlands 2005-2006", Ministry of Economic Affairs in association with the Ministry of Education, Culture and Science in the Netherlands and Ministry of the Interior and Kingdom Relations, at: http://minez.nl/dsc?c=getobject&s=obj&objectid=143714&!dsname=EZInternet&isapidir=/gvisapi/.

[97] "Pipe Dreams? Prospects for next generation broadband deployment in the UK", Broadband Stakeholder Group, 16 April 2007, at: www.broadbanduk.org/component/option,com_docman/task,doc_download/Itemid,7/gid,930/.

[98] "Rethinking Universal Service for a Next-Generation Network Environment", OECD DSTI/ICCP/TISP(2005)5/FINAL, 18 April 2006 at: www.oecd.org/dataoecd/59/48/36503873.pdf.

[99] Telecoms: the current "universal service" safety net of the EU still works well, say stakeholders", European Commission Press Release, 11 April 2006.

[100] "Broadband Deployment Is Extensive throughout the United States, but It Is Difficult to Assess the Extent of Deployment Gaps in Rural Areas", United States Government Accountability Office, 5 May 2006 at: www.gao.gov/new.items/d06426.pdf

[101] "Broadband in the universal service", OFCOM – Switzerland, 13 September 2006 at: www.bakom.ch/dokumentation/medieninformationen/00471/index.html?lang=en&msg-id=7308.

[102] "Broadband Challenge Update Dec 06: What is happening now?", New Zealand Digital Stragety, accessed on 21 August 2007 at: www.digitalstrategy.govt.nz/templates/Page____928.aspx.

[103] "An Information Society for All", Norwegian Ministry of Government Administration and Reform, 2006, at: www.regjeringen.no/en/dep/fad/Documents/Government-propositions-and-reports-/Reports-to-the-Storting-white-papers/20062007/Report-No-17-2006---2007-to-the-Storting.html?id=441497.

[104] Informatization White Paper 2006", National Computerization Agency – Korea, September 2006, at: www.dynamicitkorea.org/koreait_policy/stat_info_9638.jsp?currentPage=1.

[105] "Dempsey Unveils New National Broadband Scheme", Department of Communications, Energy and Natural Resources – Ireland, 2 May 2007, at: www.dcmnr.gov.ie/Press+Releases/Dempsey+Unveils+New+National+Broadband+Scheme.htm.

106 "Bridging the Broadband Gap: Benefits of broadband for rural areas and less developed regions, Spanish Ministry of Industry, Tourism and Commerce, 15th May 2007, at: www.bandaancha.es/NR/rdonlyres/3D83DDE8-25BB-4FC9-AA57-985F1C0F1A98/0/StoryBoardEBAen.doc.

107 "Connect Australia", Department of Communications, Information Technology and the Arts – Australia, accessed on 01 September 2007 at: www.dcita.gov.au/communications_for_business/funding_programs__and__support/connect_australia.

108 "Information about OPEL", Optus Media Release, 18 June 2007, at: www.optus.com.au/portal/site/aboutoptus/menuitem.cfa0247099a6f722d0b61a10 8c8ac7a0/?vgnextoid=e0282ad29cf63110VgnVCM10000029867c0aRCRD.

109 "The National Broadband Access Policy", Czech Republic Ministry of Informatics, 2005, at: www.micr.cz/files/2185/MICR_brozura_en.pdf.

110 "Broadband Electronic Communications in Hungary", Ministry of Informatics and Communications – Hungary, at: www.itktb.hu/resource.aspx?ResourceID=Broadband_Electronic_Communications_in_Hungary_V1.

111 www.bilgitoplumu.gov.tr/eng/docs/Information_Society_Strategy.pdf

112 www.bilgitoplumu.gov.tr/eng/docs/Action_Plan.pdf.

Chapter 2

BROADBAND DIFFUSION AND USAGE

Diffusion and usage: Market trends and developments since 2003

The value of broadband lies in the services it provides the applications it facilitates, and the content that can be accessed. What is also important is how users embrace this technology, and the impacts of expanded supply and use. Clearly, the relationship between the development of broadband infrastructure and the rollout of new broadband content and services (*e.g.* online government services) is a dynamic and interactive one.[113]

Broadband usage in businesses, schools, households and broadband applications of different sorts has grown substantially. The following sections point to significant progress in many areas but also highlight the challenges that remain in this area. When it comes to advanced broadband uses and applications, OECD countries may well only be at the very beginning. The full potential of broadband and its possible applications, in fields not related to entertainment, has not yet been fully exploited.

Business use of broadband

Following rapid growth between 2003 and 2006, broadband is now widely diffused in enterprises in OECD countries in most if not all enterprise size classes (Figure 2.1). Japan, Iceland, Sweden and Finland had the highest proportion of firms equipped with broadband, all three having roughly 90% or more firms with more than ten employees connected.[114] When only considering the EU15, penetration increased from 40% in 2003 to 65% in 2005, which amounts to a growth of more than 60% in two years.[115]

In a number of OECD countries, however, the level of broadband diffusion to firms (whether large or small) is still far from reaching the high rates of leading countries.

Figure 2.1. Business use of broadband, 2003 and 2006

As a percentage of businesses with 10 or more employees[116]

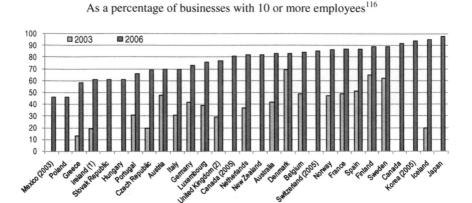

1. Includes all of NACE 92.

2. Includes all of NACE 55.

Source: OECD, Eurostat EU Community Survey on ICT usage in enterprises, Korean National Information Society Agency and Japan MIC, Communications Usage Trend Survey.

Furthermore, differences in broadband access between larger enterprises and smaller ones persist. Even in high-access countries such as Korea very small firms (fewer than four to ten employees) have considerably less access (32% for firms with less than four employees and 76% with fewer than ten employees) than larger firms which have an estimated 100% access.[117] Whereas in 2005 large Canadian firms have embraced high-speed Internet universally, about 79% of small firms with fewer than 20 employees were using broadband Internet. In the UK nearly 99% of businesses with 1 000 or more employees had broadband in 2005 compared to 70.8% of those with 10 to 49 employees.

However, in most OECD countries the growth of broadband access has been fastest for small firms (less than ten employees or 10-49) which started from lower access levels. For example, in spring 2006, 83% of all Finnish enterprises with at least five employees had broadband connection, whereas in 2003 the proportion was 54 %.[118] The broadband penetration of Australian small and medium-sized enterprises (SME) increased to 91% in 2007 (up from 63% in June 2005).[119]

E-commerce and e-business

Buying and selling online has steadily increased since 2002/2003 and broadband users are more likely to be using the Internet for e-commerce.[120] For example, broadband users in the United States are 20% more likely to purchase online than narrowband users.[121]

In the United Kingdom the retail value of e-commerce has increased by more than five times and in Australia by more than three and a half times. The total volume for online shopping in Korea was recorded at KWR 10.7 trillion in 2005, increasing by 37.4% from 2004.[122] Almost 7 million Canadians aged 18 and older placed an order on line in 2005 and Canadian e-commerce sales have seen the fifth consecutive year of double-digit growth in 2006.[123]

Yet, expectations of growth in e-commerce have not been reached. E-commerce still represents a small share of total retail sales.[124] In the area of business-to-consumer transactions there is large scope for new entrants and new players to emerge, as current transactions are concentrated with a few leading players. Around 70-95% of all e-commerce transactions are conducted between businesses. Also, cross-border transactions are still the exception. Around 75% to 90% of all e-commerce transactions in OECD countries are still national in scope. This is due to a number of problems relating to consumer confidence, online payment issues, and other barriers.

E-business applications are developing fast. Firms using broadband are more likely to be conducting modern electronic business with supply chain integration with suppliers and customers, and other modern IT-enabled management techniques.[125] Large sales in systems software and e-business applications indicate that businesses are adopting new and more mature e-business solutions, even if these investments may still be limited to large companies or early adopters.[126]

School and university use of broadband

Access to ICTs in education has steadily improved in OECD countries, with PC to student ratios up and more and more schools and universities having broadband (see Figure 2.2). But there is a general tendency for lower school levels to have less broadband access than higher levels; they are also more likely to have older equipment.[127] In most, if not all, OECD countries students still find more opportunities to access a computer and to connect to the Internet at home than at school.

Figure 2.2. Broadband in primary and secondary schools, 2006, or available year

As a percentage of schools

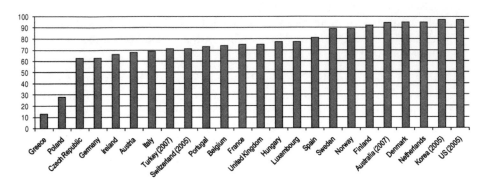

Source: National Statistical offices or regulators.[128] For EU members: EU report on 'Benchmarking access and use of ICT in European Schools 2006'. For US: Department of Education. For Australia: Ministerial Council on Education, Employment, Training and Youth Affairs.

In the US, 97% of public schools used broadband connections to access the Internet – *i.e.* 94% of small schools compared with nearly 100% of large schools.[129] In 2007, nearly all Australian government schools had broadband, with 69% having access to 2 to 10Mbps connections. Canada reports that already in 2004, 86% of schools used broadband to access the Internet.[130] In the EU, about two-thirds of schools benefit from broadband access.[131] In Turkey, in 2007, over 29 000 public schools out of 41 000 (around 70%) had access to broadband.

Despite the generally high levels of connectivity among schools in some OECD countries, there are still important variations, ranging from about 90% of schools in Scandinavian countries and in the Netherlands, to under 35% or much less in others. Schools in densely populated areas are more likely to have broadband than those in less-densely-populated areas. In Australia, for example, metropolitan government schools continue to have greater access to 2 to 10Mbps connections (76%) compared to schools in provincial cities (68%), schools in provincial areas (67%), remote schools (45%) and very remote schools (16%), although the gap between these schools is narrowing.[132]

There are no reliable statistics on the spread of broadband in higher education institutions. One plausible hypothesis is that in some OECD countries, growth in this area mimics the growth in schools, with higher education institutions often providing broadband access points such as Wi-Fi. In the United States for instance, the percentage of Wi-Fi enabled classrooms in public universities and colleges rose from 34% in 2004 up to 58% in 2006.[133]

Figure 2.3. Households' use of broadband[1] in selected OECD countries, 2003 and 2006 (or available years)

Percentage of households with broadband

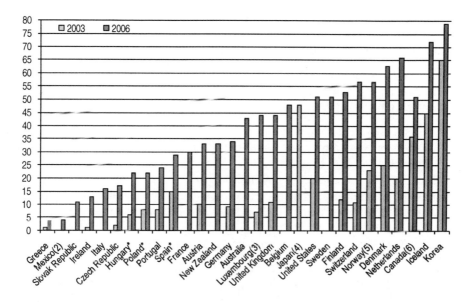

1. Generally, data from the EU Community Survey on household use of ICT, which covers EU countries plus Iceland, Norway and Turkey, relate to the first quarter of the reference year. For the Czech Republic, data relate to the fourth quarter of the reference year. Data are for 2004 rather than 2003 for countries marked with an asterisk * (*e.g.* Czech Republic). For the Czech Republic and Canada the last value corresponds to 2005. The value from the Australian Bureau of Statistics is for the years 2006/2007 and for 2007 for the United States.

2. For 2001 and 2002, households with Internet access via cable. From 2004, households with Internet access via cable, ADSL or fixed wireless.

3. For 2004, data include wireless access.

4. Only broadband access via a computer.

5. For 2003, data include LAN (wireless or cable).

6. Statistics for 2001 and every other year thereafter include the territories (Northwest Territories, Yukon Territory and Nunavut). For the even years, statistics include the 10 provinces only.

Source: OECD, Eurostat EU Community Survey on ICT usage in households, Korean National Information Society Agency and Japan MIC, Communications Usage Trend Survey (figures are rounded). This household data does not take into account non-subscriber access to broadband services.

Broadband access and use in OECD households

Broadband has spread much faster to households than the telephone, the computer or similar ICTs. Whereas a few years ago broadband usage and access was confined to a small group of individuals, it has now increased significantly since 2003 and is often close to universal (see Figure 2.3). In 2006, nearly 80% of all Korean households used broadband at home.[134] Iceland had a proportion of 72%, followed by high shares in the Netherlands (66%) and the Nordic EU countries. Those OECD countries with originally lower access levels have often experienced fast growth rates which caught up with the more advanced countries.

Nevertheless, only nine OECD countries have 50% or more of households using broadband and some OECD countries are still facing very low household uptake of broadband.

In OECD economies, wireless connectivity of PCs, laptops via Wi-Fi hotspots or private routers in households and firms is becoming more widespread. Of the 21% of adults in the UK with a Wi-Fi-enabled laptop, one in three has used it to access the Internet while away from home. In the US, 34% of Internet users have logged on with a wireless Internet connection either at home or at work.[135] Municipal Wi-Fi networks, new services which allow homes and small business to operate Wi-Fi networks and other Wi-Fi offers (*e.g.* Fon.com – discussed earlier) are improving wireless connectivity.

While official data on the existence of Wi-Fi hotspots does not exist,[136] more and more "hotspots" have appeared since 2003. Figure 2.4 shows the number of public wireless hotspots as provided by a private consultancy. According to this data, the United States alone presently represents one-third of the world's wireless hotspots. Despite continuing growth, however, the numbers relative to population are still low in OECD countries.

Broadband is not only leading to an increase in the share of the population conducting online activities, but also to more frequent and longer use. For instance in France in 2006, among broadband connected people, seven out of ten were accessing the Internet daily, compared with three out of ten people who are on narrowband. The United Kingdom had almost 30% more broadband users accessing the Internet daily compared with narrowband users.

Before, the most common activities on the Internet entailed accessing news and information, receiving e-mail and instant-messaging, conducting simple online banking, and online shopping.

Figure 2.4. Public wireless hotspots per 100 000 population, fourth quarter 2006

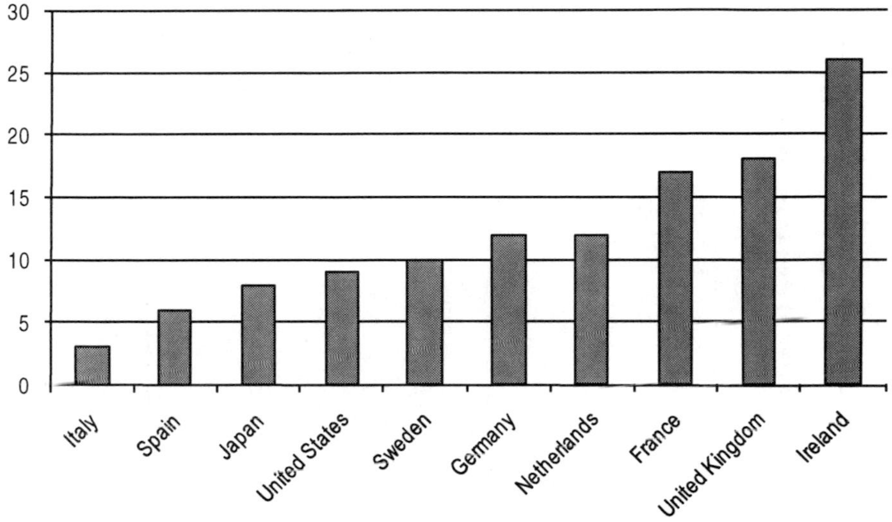

Source: Informa telecoms and media.

More recently, new content-rich broadband applications and new forms of usage have become a key driver of broadband demand and uptake. The availability of broadband has *reinforced existing activities* (*e.g.* e-mail, news and information, shopping online), but this has also *brought about new forms of usage and innovation* (*e.g.* video streaming, podcasts, high-definition television over broadband). In 2006, for example, listening to radio or watching TV via the Internet has gained in popularity – see Figure 2.5. In particular, developments in 2006 and 2007 brought about an impressive array of new media *(i.e.* launch of online content services with interesting titles and pricing schemes and portability). Other non-entertainment forms of broadband use are still emerging (*e.g.* in the area of health). Broadband has also enabled persons (including the elderly, the sick, and, in particular, rural inhabitants) to stay in touch.

Figure 2.5. Internet users listening to web radios/watching TV, selected OECD countries, 2002-2006

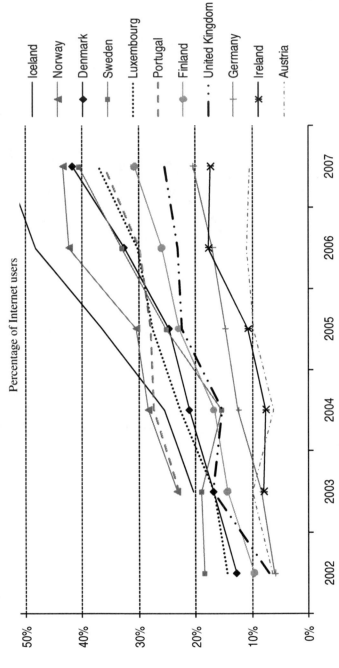

Percentage of Internet users

Source: OECD Information Technology Outlook 2008. Based on Eurostat data.

Broadband users are also more participative.[137] They contribute content to web sites, they keep online diaries and blogs, and share photos, videos and artwork. For example, in the United States, more than half of all teenagers aged 12-17 were involved in online activities in 2007, and many of these were actively creating content.[138] In fact, the limits currently placed on broadband up-stream traffic are significantly slowing down participation rates. With current activity, and as new types of broadband uses emerge, insufficient broadband access is leading to serious usage divides, which warrant policy attention.

However, there is evidence that different socio-economic groups have access to and use the Internet differently.[139] Despite progress in broadband usage and access, certain divides are evident. Household use is often related to income, education levels, gender (males having more access),[140] the number of children (households with children having more access), age and disability. As data for 2006-2007 from Australia shows, use is significantly higher for the age group 15 to 17; people from households in the top two income quintiles; people with higher levels of educational attainment; and the employed.[141]

To give specific examples relating to age groups: while broadband use is almost universal among certain young age brackets, people over 60, 70 and older are less connected, contributing to an increasing divergence between generations mainly due to an absence of PCs in these households. In the US, 15% of the population (up from 8% in 2005) over 65 years had access to broadband in 2007 versus 63% of those in the 18-29 age bracket.[142] In the UK, in 2006, 84% of people aged 16 to 24 years had used the Internet within the last three months, compared with 52% of people aged 55 to 64, and 15% of those aged 65 and over. In Korea, the broadband usage rate of persons in their 60s and older is around 16.5% (see Figure 2.6). In Japan, generational disparities generally lower with respect to mobile broadband access.[143]

Yet broadband access is growing fastest for those without previous access. In spring 2006, for example, Finnish Internet use among 60 to 74 year-olds stood at 36%. The corresponding figure was 27% one year earlier.[144] In Germany, the growth rate of Internet use among the age bracket of 50-59 years is also slightly above the country average, at a use rate of 34% in 2006, compared with the country average of 58%.[145]

Figure 2.6. Korean broadband use, December 2005 to December 2006, by age group

In percent of total population of age group

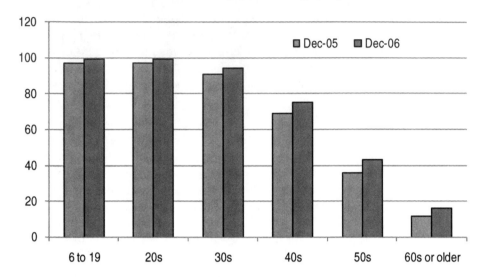

Source: NIDA, Korea

Broadband is also increasingly accessible to very young age groups in some OECD countries. In Korea more than half of all three to five years-old are using broadband at the end of 2006.

The section on infrastructure has also shown that although rural areas have seen substantial broadband growth[146], users in these areas often face a significantly lower quality of access. Concerns over households or individuals who are "digitally excluded" remain. As broadband becomes more and more important in daily life, the disadvantages of those excluded will be accentuated.

Increased range of broadband applications and content since 2003

The role of broadband as a platform for communication and content distribution has evolved significantly since 2003. Convergence is bringing new opportunities for businesses and consumers. While in 2003 digital broadband content services hardly existed or had a small user base, in 2007 new services such as Voice over Internet protocol (VoIP), online music platforms, online gaming and other innovative interactive services, equipped with an always-on functionality and providing broad bandwidth capacity, have been put in place. Even higher data-intensive applications are on the horizon, *e.g.* streaming high-definition video, new peer-to-peer applications, high-definition television downloads, health or education applications, virtual conferencing, virtual reality applications. With the rise of the concurrent use of such online applications, bandwidth requirements are on the increase.

The main drivers for this emergence of new broadband uses are: technological developments (mainly content protection technologies, portable media players and technologies which allow for the shifting of content from the PC to the TV); widespread broadband access and wireless networks; increased use of broadband in business-to-business relations; greater uptake and demand ; a greater willingness to pay for Internet content; greater confidence; lower entry barriers (*e.g.* for content distribution); and new business models. Through broadband, Internet use has also become more participative as users increasingly upload their own content, which also makes upstream bandwidth more important.

Even in 2007, however, this evolution towards more broadband applications and increased use is far from complete. Suppliers of broadband services and digital content have to reinvent their technology and services as well as the delivery of these services. In fact, compared with earlier estimations, new broadband applications such as converged digital content services (online music, film and video, etc.) took longer than expected to arrive. Digital content services, video-on-demand, video conferencing and interactive offers are still in their experimentation phase. The goal of "broadband content anywhere, anytime and on any device" is still remote. The widespread use of advanced mobile broadband services has yet to emerge in most OECD countries.

There are multiple challenges to broadband development which need addressing:[147]

* Availability, pricing, and speed (*e.g.* for larger high definition video), quality of service and other technical issues (discussed earlier in terms of supply).

* Organising broadband applications and services is a substantial challenge. The introduction of broadband often poses serious challenges to pre-existing business models. The development of digital content services requires co-ordination of a wide range of industry participants, some of whom have not previously worked together, thus taking some time as existing business models are being challenged.

Some of the key discussions among players in the value chain revolve around the topic of revenue sharing (*i.e.* mainly between telecommunications operators who carry risk and investment in developing advanced infrastructure and content and other application providers). Broadband operators face uncertainty as to how to recoup their large investments in the absence of new revenue-generating broadband services and content. Content providers are waiting for improved connectivity and content protection.[148] These mutual uncertainties have the potential to slow down investment in higher-speed broadband networks and the generation of new broadband services.

Other bottlenecks include: digital piracy; interoperability at various layers (*i.e.* digital services and content working on certain devices only, services or network connections); open content delivery models versus walled garden approaches (increasingly in the mobile environment); rights issues such as the difficulty of securing rights and accessing content for global Internet markets due to the territorial nature of copyright; existing exclusive distribution deals; conflicting rights; windowing; and the absence of innovative solutions to exploit and license content online.[149] Fragmented standards (for mobile platforms and digital rights management), the absence of online payment systems, regulatory issues (regulation of non-linear services vs. linear ones), competition aspects and various other obstacles (including access to finance, skills) make progress difficult.[150]

Triple play services

Since 2005, most OECD countries have seen an increasing availability of bundled communication service offers. More and more Internet Service Providers are offering so-called triple play services (voice, Internet and film/video). An analysis of 87 providers in the 30 OECD countries found that multiple play offers of video, voice and Internet access (triple play) were available from 48 providers in 23 OECD countries in September 2005.[151] Today, the majority of telecommunication operators have moved closer to becoming all-in-one service providers for voice, video and data.

In the past three years there has been a large increase in the number of multiple play subscriptions (triple play or quadruple play) which package together video, voice and data. In the United States, for instance, competition between the cable and telecommunications companies has driven much of the market towards offering triple-play services, and increasingly, wireless services are becoming part of that bundle. These offers attract consumers because they offer a simple, consolidated bill and are typically less expensive than if the consumer bought all the services separately.

Today some triple play packages in France, for instance, offer 28 Mbit/s downstream and 1 Mbit/s upstream, access to around 250 TV channels (among which 100 free and the other 150 optional), and free national voice calls and to fixed lines in 70 foreign destinations at a price of EUR 29.99. Such triple play offers in OECD countries are constantly being improved with faster transmission speeds, more TV channels and on-demand video content and more destinations to which consumers can call for free over VoIP.

Video-on-demand over broadband

Market analysis and data on penetration levels show that the video-on-demand (VoD) market is growing quickly.[152] Apart from broadband speeds, the improved availability of VoD content with more favourable access conditions, and the emergence of high-definition films and television have accelerated this trend. However, in terms of size, most online on-demand video markets in OECD countries are still small and developing.[153]

Services for downloading movies are on the rise in OECD countries. However, these services have not yet gained significant commercial traction. There are also problems with different formats, limited catalogues, offers restricted to certain national territories, and problems related to payment.[154] Most of the interesting offers with a rich content-catalogue are still only accessible in countries such as the United States, and integrated catalogues (*e.g.* of recent European films accessible in all European countries) are not available at all. Several on-demand video download services have emerged – but they are usually only national in scope and often only in their first months of operation (*e.g.* Apple).[155] Terrestrial broadcasters have started to offer live streaming of their content or have VoD services on offer (*e.g.* Arte, CanalPlus) but all in all the offers are very limited.

Recently, however, social networking sites such as YouTube and MySpace have been more successful in attracting users than the conventional video download offers. As ISPs innovate through selling VOD services as part of their broadband package and as content rights are increasingly secured, the online VoD markets will develop at a faster pace. New innovative peer-to-peer based services (*e.g.* BitTorrent and Joost) will drive online video formats further.

Internet Protocol Television

Internet Protocol Television (IPTV) offers – which are different from VoD as defined by the OECD[156] – are also at an early stage. Available figures show the limited penetration of IPTV offers (see Figure 2.7). Only in a few European countries such as Spain, France, Ireland and Italy are IPTV services starting to become a mass-market service due to triple play offers.[157]

Figure 2.7. Internet protocol television penetration, December 2005

Subscriptions per 100 households and DSL connections

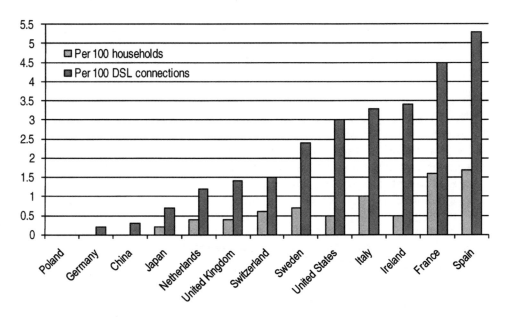

Source: Ofcom, The international Communication Market 2006[158] based on national regulators, OPTA, and operators. Figures for Switzerland are for 2006 and have been provided by Bakom

Recent analysis, however, shows that IPTV subscriber numbers have grown fast in the last year, effectively doubling during the 12 months to 30 June 2006.[159] Fibre, high definition IPTV, three-dimensional TV, video applications for tele-work and tele-medicine will push demand for high bandwidth further. Eventually this demand for heavy video files will put pressure on the existing capacity of the Internet.

Use of mobile applications and mobile content

Mobile content has long been viewed as a major driver of growth in the telecommunications and media industries. Higher speed data transfers, more affordable, better handsets and increased availability of content for the mobile market are elements driving growth.[160]

Despite this recent growth in mobile content, and despite existing 3G networks and services, in most OECD countries (except for Japan and Korea) the uptake of mobile content services is taking longer than expected. Such technology is at an early stage, with most users in OECD countries only sending SMS messages, and buying ring tones, music downloads and simple games. Access to mobile online content will likely improve as operators upgrade networks to support higher-bandwidth data services and adapt their business models. The success of these higher-bandwidth content services will hinge on competitive pricing and usage plans promoted by operators. High data-variable costs, unclear pricing schemes, site load times, high roaming costs, navigation problems, user-unfriendliness and lack of demand have been problems in the past, which have also led to a discrepancy between web access possibility and actual usage. It remains a challenge to adapt and package content to suit mobile platforms and to increase interactivity and sharing.

Improved bandwidth on mobile networks could help operators exploit online media. If users had access to higher-bandwidth and inexpensive connectivity via their mobile handsets, governments could also promote increased public sector content usage such as such as health information, educational materials and other government produced digital content. [161]

Advanced broadband applications for government, education, health and other social sectors

While entertainment-related applications and uses have played a key role in broadband uptake so far, broadband is also very important for areas of high government interest. These include tele-work, health, energy, education and government services.

Since 2003, advanced broadband applications of this type have been set up and pilot projects have developed. Yet, OECD firms and countries still have a long way to go to seize the full potential of broadband in these areas.

Tele-work

Tele-work has the potential to reduce pollution and government transportation spending, to increase the efficiency of the workforce and to allow for a better coupling of family life and work.[162] Available data point to an increase in tele-work practices in OECD countries. In 2006, 19% of all workers – *i.e.* 28.7 million US employees – tele-work at least one business day per month, an increase of 63% from 2004 to 2006.[163] About 1.5 million persons tele-work in Canada.[164] In 2006, over one-third of Australian SMEs introduced tele-working, with broadband being a determining factor.[165] In Europe, the differences between non-broadband and broadband enterprises with regard to tele-work are very pronounced: 2.5 times as many broadband enterprises offer tele-work as non-broadband enterprises in Europe.[166]

Yet available data and experience show that the opportunities of tele-work are not fully exploited. In 2006, a study conducted by the Finnish Ministry of Labour, for example, showed that only 1% to 4% of all wage earners were engaged in telecommuting, as opposed to 14% to 19% of all wage earners who could conceivably tele-work. A similar picture emerges for most other OECD countries. There are still barriers to the development of tele-work, namely inflexible organisational structures and cultures, lacking infrastructure, and rules that do not allow for tele-work.

At the same time, although tele-work may offer many advantages, tele-work policies should also take into account negative aspects such as 'rebound effects'.[167]

Advanced broadband applications and (e-) government

Broadband has the potential to streamline the internal workings of government, facilitate the on-line supply of public services and provide access to public sector information and content. Statistics show that broadband users are more likely to interact with public authorities over the Internet and their websites than narrowband users.[168]

The public sector has increasingly used the Internet as a basic communications tool for many government activities, from simply providing information to supplying fully transactional services integrating several public services, which forces governments to rethink business processes and service delivery.[169] OECD countries such as Canada, the United States, Denmark and Australia have been leading the way.[170]

Many government institutions in OECD countries use the Internet on both federal and regional levels to communicate information. In Japan, for instance, all federal, regional and local government organisations have their own websites.[171] These e-government services include online taxation and online access to legislation and regulations in different countries.[172]

With *always-on*, high-bandwidth networks, interactions with business online are also becoming more sophisticated with some OECD countries offering one-stop platforms for government procurement, bidding information, etc. E-services requiring high volume transmission capacity for improving openness and access to democratic institutions are becoming feasible. Examples include Internet broadcasts of parliamentary debates, or the use of multi-media content within the educational or cultural sector.

Despite progress in the use of broadband by governments and citizens, there is room for significant improvement. Considering the high availability of broadband and – at times – e-government services, the take-up of such services often remains low, even for leading OECD countries.[173] Country studies on e-government in a number of OECD countries also show that the largest challenge is still the organisation of back-office processes.[174] A lot of potential also exists in areas such as public safety, *e.g.* putting in place wireless broadband networks with mobile high-speed data, video and audio feeds to police, fire fighters and emergency crews.

The barriers to broadband innovation in government still include resistance to change and the restructuring which is necessary to successfully deliver shared services; antipathy towards data sharing; general aversion to risk; and the inability to clearly communicate and demonstrate the benefits of e-government services. Informing citizens about services[175] and creating single entry points are key.[176]

Advanced broadband applications and education

Broadband and distance education are important elements in modern education systems which help to improve life-long learning, and the quality of the workforce. Moreover, Internet access can help make educational online activities more attractive. Indeed, there is evidence of improved educational performance through the use of ICTs in various studies of the OECD.[177]

Broadband stimulates the use of the Internet for educational purposes. For instance, in Denmark in 2005, a quarter of all users in broadband households used it for educational activities, whereas only 14% of users in non-broadband households used the Internet for educational activities.[178] Data for the EU also shows that there is a clear relationship between the percentage of teachers using ICT in teaching and the percentage of schools with broadband connections in European countries.

Access problems and achieving necessary bandwidths are still ongoing challenges.[179] A number of OECD broadband policy reports note that despite positive efforts, the education sector is still characterised by differing levels of connectivity.[180] In the future, connections must offer significant speeds to support applications such as collaborative workspaces, distance lectures, videoconferencing and the like.[181]

Beyond access, often the real challenge is to create the right incentives for teachers and students to use these broadband applications in the educational setting. Efficient use depends on computer to student ratios, ICT skills, incentives for teachers, and educational applications. How broadband applications should be applied to the schooling for tomorrow is a question that needs to be addressed rather sooner than later.

Advanced broadband applications, health and ageing

The potential for broadband-based medical applications and other economic and social benefits of broadband for health and ageing is large.[182] One study asserts that remote patient monitoring could cut health care expenditures by 25% in the United States.[183] Studies for Europe echo that the cumulative benefits of ICT projects in health exceed the costs by far.[184] Broadband can also help older adults to live active, independent lives.

More and more broadband users are taking responsibility for managing their health by consulting medical information and over broadband networks and applications.[185] Discussion platforms on social networks and health broadband TV channels on online platforms such as Joost feature health videos, real-life patient experiences, etc.

However, broadband has the potential to do more. E-health applications could help with remote consultations (*e.g.* monitor heart patients), improve the links between caregivers, health institutions and patients (*e.g.* sharing of X-ray images),[186] to more sophisticated applications such as tele-medicine. Electronic patient records and other digital paper trails could enhance the workings of health institutions as well as speed up processes.

Originally, the major impediment to the more widespread use of broadband in the health sector was availability at an affordable price. Today, it is mainly rural areas that face access issues. In addition, the tele-health applications will require more bandwidth than currently available.

While progress has been made on the access front, the rate of adoption of these health-related applications has often been very slow.[187] Many of the above-mentioned examples are not widely diffused in OECD economies. Finally, although various healthcare activities and functions make intensive use of ICTs (*e.g.* imaging technologies), there is great potential for further integrating digital delivery and interaction, *i.e.* in care, research, teaching and evaluation.[188]

Impediments are primarily human and organisational rather than technical. Beyond infrastructure, institutional rules and practices that foster the use of broadband need to be introduced. Skills issues, the interoperability of systems, security and privacy concerns and other barriers need to be addressed.

Other advanced broadband applications for other business and social sectors

Since the emergence of the Internet, broadband applications were envisioned which would permit the remote delivery of services geared to increase the everyday quality of life and provide solutions to key policy problems. The home of tomorrow was envisioned with smart applications which would facilitate every day activities (*e.g.* refrigerator automatically ordering certain products when running low) or which would facilitate the rise of smart automated homes permitting the remote control of energy, security, etc. Further down the road, a ubiquitous network society of services based on 'person-to-person', 'person-to-goods' and 'goods-to-goods' communications (*i.e.* the Internet of things) was envisioned. The latter would help providing solutions to in the area of traffic management, carbon reduction, care of the elderly, etc.

Since 2003, trials have been ongoing in these areas. In practice, however, the development of these broadband applications is still a challenge and a matter for current and future work.

Diffusion and usage: Application of the Recommendation

The most advanced OECD countries focus on increasing uptake of installed capacity and paying more attention to electronic business, digital delivery and broadband applications. Promoting the general ICT policy environment, fostering innovation in ICT, ICT diffusion and use (including *e*-government), ICT skills and employment, digital content, the ICT business environment, and promoting trust are important components of these policies. The growth of broadband use also requires a number of interventions going beyond ICT policy, such as reform of the public sector, and new approaches in research and education.[189]

Demand-based policy approaches

The Recommendation emphasises the role of demand-based approaches for broadband development. Policies are encouraged which promote access to all communities, which encourage investment in infrastructure, content, and services and which also promote take-up and use. These demand-side measures cover a very wide range of ICT policies, some of which are specific to broadband and some that are not, making the activity more diverse, more difficult to co-ordinate and measure policy impacts. At the same time, the impact of some of these policies is potentially more far-reaching (*e.g.* e-government and related reforms).

Raising awareness of the advantages of broadband

Many OECD countries have initiated broadband awareness and campaigns (*e.g.* seminars, road show events, conferences, presentations of best practices) to promote the benefits of broadband. These campaigns highlight the applications, content, and the advantages of broadband. Sometimes they aim to help citizens choose among different broadband providers.

For instance, the Australian government took measures to assist individuals and groups to make informed decisions about the adoption of broadband. As part of its Plan for the Development of Broadband Services until 2008, the Greek government recently established 85 access points with information about broadband benefits. Turkey is running pilot project for ICT awareness and usage in rural areas.[190] In the United States, ConnectedNation, a public-private partnership, in conjunction with affiliated state programmes have analysed both supply and demand issues to help increase broadband availability and usage as well as bridge the digital divide.[191]

Other campaigns are concerned with promoting future broadband services (see, for example, Box 2.1).[192]

> **Box 2.1. Nederland BreedbandLand (NBL)**
>
> Nederland BreedbandLand (NBL) is an independent national platform for the provision of aid and incentives to the social sectors (*e.g.* healthcare, SMEs, government, transport) for the 'better and smarter' use of broadband. The NBL identifies and accumulates knowledge and experience from broadband projects and bundles it in a knowledge database. It disseminates this knowledge through seminars, workshops and other related events. In addition, the NBL tracks potential 'breakthrough' community broadband projects on a sector-by-sector basis to support their development into best practices and to ensure expansion to national level. New directives and standards may emerge from the knowledge. Other countries such as Belgium have been emulating such projects.
>
> *Source*: www.nederlandbreedbandland.nl

Policies to improve broadband access for businesses

Diffusion of broadband to businesses has been at the core of broadband policies in many countries for years. Today, such policies are no longer really needed in many OECD countries. These days, fostering broadband diffusion to businesses is mainly centred on SMEs and start-ups and encouraging the use of broadband applications. There is also a focus on skills formation, demonstrations of best practices and the benefits of use.

Governments, academia and the private sector, on either an industry-specific or economy-wide basis, have set up fora to share knowledge on communications technology. These fora are often centred on specific sectors or topics such as network security and the adoption of e-business by SMEs. Information campaigns about the benefits of e-business are increasingly the focus. Awards for excellence in e-business or related technology-based innovations exist, as well as initiatives to promote e-procurement, e-invoicing, etc.

The programmes aimed at SMEs focus on promoting the benefits of broadband, addressing the shortage of appropriate skills and applications, showing how broadband can help access to government procurement and other means to stimulate demand. Certain OECD countries are envisaging specific financial incentives, such as tax credits for SMEs which invest in ICT assets and broadband connectivity. Other government actions are concerned with creating business districts with very fast broadband access (*e.g.* French "Zones d'activité très haut debit"). Helping businesses with necessary skills, to adapt their organisation and business models to broadband (*e.g.* Finnish National Broadband Strategy), and to help with broadband standards (*e.g.* German PROZEUS project to settle on e-business standards) are popular.

Renewed efforts may be needed in promoting the effective use of broadband in the case of businesses that have access. There is still significant potential for using more advanced broadband services and for e-commerce and e-business applications.

As mentioned previously, the relatively low uptake of tele-work is also an issue. Many OECD broadband policies recognise the potential of tele-work, and while some have explicit numerical targets (*e.g.* Japan, plans to increase tele-workers to 20% of the working population by the year 2010), across the board there are few practical government initiatives or incentives to foster telecommuting. Finland, one of the exceptions, however, has adapted national rules to make mobile, tele-working and other flexible work more feasible.

Connecting schools, hospitals and other public institutions

Many governments are seeking opportunities to aggregate broadband demand across health and education and other public institutions to support the rollout of broadband infrastructure in regional areas. These policy plans are complementary to supply-side initiatives to increase broadband deployment. The UK has compiled a best practise guide for public broadband schemes.[193]

To take the example of schools and the educational sector, most OECD governments have granted subsidies for the purchase of ICT connectivity for schools and research institutions. In New Zealand, the Project PROBE has been developed to rollout broadband, *i.e.* in particularly to remote schools and provincial communities.[194] In the United States, the Schools and Libraries Program of the Universal Service Fund gives discounts to eligible schools and libraries for Internet access.[195] In Turkey, schools without access, which are located in geographically difficult areas, will be connected to the network using alternative wireless access techniques.

Many government efforts also target other institutions. Connect Australia, for instance, includes the Clever Networks programme to provide high-speed links for our universities, hospitals and research institutions, especially in rural and remote areas. Japan and Korea are particularly interested in putting the network infrastructure to use for social welfare, nursing, and in other public institutions.[196]

These above measures have significantly increased the number of connections. As described in earlier sections, however, the level of connectivity in schools, hospitals and similar institutions still varies widely within and among OECD countries. Many of the reviewed broadband and wider ICT policy plans state the goal of improved connectivity in these institutions. Yet such policy goals, often expressed in percentages to be achieved (*e.g.* 100% of schools connected by 2006), need to be monitored and pursued in earnest.[197] Too often, such policy goals are not pursued efficiently and in an accountable manner.

Policies to improve broadband access for households and individuals

A large majority of OECD countries have introduced initiatives to diffuse technology to individuals and households. While some years ago, policies focused on increasing basic ICT penetration at home, work, and school, now policies are more focussed on broadband reaching specific hard-to-reach groups and on stimulating demand through the availability of broadband content.

Some countries consider the stimulation of PC penetration as an important element of broadband national strategies to foster Internet use. This entails fiscal incentives to purchase a PC and/or access to residential and business broadband access (*e.g.* Italian "PC for Young Scheme" and similar schemes in Austria and Sweden). Belgium is providing incentives for broadband consortia to offer a package of PC, software and broadband, *i.e.* citizens receive tax breaks. Aside from households, countries such as the United States fund broadband access to certain types of organisations. Specific programmes such as subsidies for teachers and the provision of recycled hardware to low-income homes complement these initiatives. Fiscal incentives to promote the teaching of computer classes at home are also being tested.

Many countries offer public broadband access points to cater to persons without access (*e.g. e*Hungary points, the Internetstuffen in Luxembourg), often in places such as libraries and community centres. These tend to target underprivileged communities and those outside of major urban centres. The EU also plans special incentives to cater to rural and remote areas.[198] Increasingly, broadband policies in OECD countries cater to specific groups such as the disabled, the unemployed, those entering the workforce, indigenous communities (especially in Australia and Canada) and the elderly and low-income households.

As shown earlier sections, basic access issues are becoming less of a problem in many OECD countries and existing policies seem to have been successful. Attention to the digital divides that remain, to the newly arising usage divides and to those lacking ICT literacy now merit further analysis and policy attention.

ICT skills

Fostering ICT skills and sustaining e-learning efforts is an important part of OECD country broadband plans and is consistently ranked high priority. These programmes also target specific groups such as women, the elderly, the unemployed or the disabled. The integration of ICT skills into education at all levels from elementary to postgraduate and in vocational education and training has also been a priority. Increasingly these programmes aim at integrating broadband and ICTs into the curricula and educational material (*e.g.* the Spanish Avanza Plan). In some countries the formation of broadband skills is part of programmes to foster media competence (*e.g.* the Swiss 2006 Information Society Strategy).[199] Denmark and Japan, for instance, are developing programmes to foster the web ethics for children and young people.

Promoting broadband content and applications

Governments can aid the development of broadband by acting as model users, not only through their role as a purchaser of services, but also as major providers and content developers in key public policy areas such as health and education.

Move to more sophisticated government online services

In OECD countries, policy initiatives to foster more sophisticated government online services are becoming popular. While many public services are now available online, most OECD governments announce numerical targets to improve them further (*e.g.* Turkey with the goal of having 70% of the public services online by 2010).[200] This includes expanding secure government networks, putting administrative processes and documents online, supplying firms and citizens with more cost-effective ways to deal with the government (including once only submission of data), and, for example, assigning firms and citizens a single number or identifier to conduct their relations with government.

Since 2005, a large number of countries specifically cite a move towards citizen-centric government (*i.e.* measuring user satisfaction, user-friendliness). This change in focus largely reflects concern about the slow take-up of e–government services under the previous government-centric models. In practice, this means a shift towards "one-stop shops" and mainstreaming e–services via integrated multi-channel service delivery strategies (*e.g.* the Norwegian Høykom strategy).[201]

Governments promoting broadband-related standards

OECD governments have implemented various standards to spur broadband use by businesses and citizens (including open standards).

* *E-procurement and electronic payment:* Government-sponsored demonstration projects or standards are used in the OECD to promote or encourage e-procurement and e-invoicing.

* *Data exchange standards:* Governments also support standards in the field of data exchange and electronic business exchanges, often as lead users.

* *Digital signatures and identifiers:* OECD governments are conducting projects that relate to electronic identification, the supply and use of digital certificates for signatures and identification. Government projects are also concerned with supplying digital identifiers.

Over the last few years, governments have had mixed success in establishing standards, digital signatures and electronic payment mechanisms. Many schemes set up in the early days of the Internet never developed or did not find a satisfactory user base and were then abandoned. New approaches have to ensure that there is sufficient market support for the chosen standard and that the introduction of these standards is well orchestrated (with appropriate hardware, service and software providers). Government-led projects also have to consider the international dimension, *i.e.* take into account an increasing number of travellers and cross-border transactions which might weaken purely national approaches if interoperability is not ensured.

Governments putting content online

Access to and commercial re-use of public sector information: Currently many OECD countries pursue policy initiatives and laws that facilitate the access and commercial re-use of public sector information (*e.g.* geographical, traffic, economic, social data). In most OECD countries, access to this type of public sector information is still imperfect due to poor transparency, hesitant public actors, burdensome procedures, costs involved, and matters relating to intellectual property rights, etc.

Increasing the accessibility of other public sector content: Providing co-ordinated access to research, cultural and educational public resources (*e.g.* from educational and research institutions, public service broadcasters, and cultural centres) is an important policy objective for OECD countries. Increasing the transparency of available material and putting in place workable government intellectual property policies are a priority. Libraries, museums, archives and other cultural heritage records are now largely digitised; the preservation of this digital material is a recurrent policy objective. Despite these efforts, substantial scope for improving access through further digitisation remains.

Spurring digital content demand: The public sector also plays a major role, both as a customer and as a channel for aggregating broadband demand. The United Kingdom, for instance, is issuing specific guidance on how to facilitate more effective procurement of broadband content by the public sector.[202] The promotion of educational broadband content such as e-learning or distance learning initiatives is also a good example.

Supporting the development and distribution of digital content

OECD analysis has shown that governments employ a wide range of direct and indirect measures that affect digital content development.[203] Since 2003, OECD governments have also been devising various stand-alone policies to make the digital content or creative industries a strategic sector for the broadband economy (*e.g.* Australian Digital Content Industry Action Agenda, the UK Digital Content Strategy).[204]

Promoting interoperability, innovation and choice

The OECD Recommendation calls for "Technologically neutral policy and regulation among competing and developing technologies to encourage interoperability, innovation and expand choice". The interoperability of broadband services and applications on various networks and platforms is of increasing importance as users ask for the same products over different platforms.[205] A range of different proprietary and incompatible formats, networks, services and consumer devices will hamper the development of online content distribution and other broadband services.

Governments usually do not have the experience and technological foresight to pre-select standards in fast-moving areas. Often initial experimentation by the market place and the formation of industry bodies (*e.g.* the Australian Communications Alliance or the Network Reliability and Interoperability Council in the United States) is needed to develop the best approaches. However, governments can provide frameworks for co-operation and can engage business, experts and standard organisations to work together to develop better standards. As part of its i2010 strategy, the government of Germany will collaborate with industry to set standards for advanced broadband applications. The Korean broadband content growth strategy also includes efforts to create digital content standards.

Sometimes *de facto* standards are established by market leaders which can then hamper competition. With vertical integration, lock-in of consumers in certain standards, and limited access to certain content, maintaining an environment where small and innovative players can compete and where consumers can make choices between various existing offers is essential. Governments can mandate a certain degree of interoperability (*e.g.* the French regulatory authority on technical protection measures) but can also use competition policy to avoid the abuse of market power.

OECD governments sometimes encourage open and interoperable standards, and/or use specific demonstration projects or workshops to promote interoperability. Belgium and other OECD governments, for instance, are promoting open document standards in their national administrations.

Unlocking the potential behind advanced broadband applications in social sectors

OECD governments have also shown interest in fostering the development of more advanced broadband applications in social sectors. Some OECD countries have formulated policy visions for a ubiquitous information society in which broadband is the underlying technology for most economic and social transactions (see the u-Japan project and the Korean IT839 Plan but also the plans for an "Internet of Things" in the German ICT policy).

Beyond these policy visions, other national initiatives exist to spread the benefits of broadband to specific social areas. Governments have set up national platforms to showcase best practices and share knowledge to enable smarter social uses of broadband. Several similar initiatives exist in other OECD countries. Certain government initiatives reinforce advanced broadband applications through R&D support (discussed later). For instance, the Korean Electronics and Telecommunications Research Institute carries out research on advancing the digital home and telematics applications.[206] In Belgium, the Interdisciplinary Institute for BroadBand Technology focuses on applications of broadband technology in social sectors.

Various OECD countries also have visions on how to implement broadband in specific government or social areas, such as health and transport.

Broadband and health care: Most OECD countries see the potential for broadband to contain costs, improve services and deliver better health outcomes in the health care sector. Many are implementing national e-health strategies (*e.g.* the Belgian *Projet de stratégie nationale en matière de cybersanté*).

OECD country initiatives have focused on five priorities:

- Improving connectivity for hospitals and doctors (*e.g.* Iceland's health-service network, France expert networks to stimulate sharing of medical dossiers, the Canadian K-Net tele-health and the Australian Broadband for Health Program connecting indigenous and rural communities).[207]

- Setting up demonstration projects with innovative e-health examples (*e.g.* the Belgian Coplintho project[208], a tailor-made network for the support of care-related aspects at home).

- Establishing common standards to exchange data (*e.g.* Canada's plans for health care information systems designed to improve patient care and health care delivery in all regions, Turkey's efforts to introduce a health information system, Japan's plans for a healthcare public key infra-structure).

- Introducing electronic record systems (*e.g.* the plans of the United States to establish electronic medical records, the Australian plan to inaugurate a national electronic record system, Germany improving its electronic health card).

- The reorganisation of basic medical care and specialist medical care (*e.g.* Japanese plans to increase operational efficiency, enhance health care safety, and to provide diagnosis and treatment information with electronic networks and RFID tags).

As stated earlier, broadband – or ICT-based solutions for *patients* are developing more slowly. The projects listed above need to be pursued in earnest. However, a co-ordinated national approach must be taken to manage different health projects and to minimise interoperability (see report of the Norwegian Hoykom project).[209] Existing efforts to implement ideas such as shared medical dossiers – although appealing in theory – have often fared badly in practise for various reasons that need further analysis.

Broadband and transport: Most OECD countries see the potential for broadband and ICTs to reduce traffic and thus decrease the chances of road accidents and. Some OECD countries have put initiatives in place that will foster the use of broadband networks for these goals. For instance, the Japanese government is supporting collaboration between the public and private sectors to put into practical use intelligent transportation systems (ITS) as a means of reducing traffic accidents, *i.e.* guided by specific numerical policy goals (*e.g.* percentage reduction of traffic fatalities). Given the high potential for these transport-related initiatives, in order for them to succeed, such projects need renewed attention and a concerted effort of different departments and levels of government, and potentially the collaboration of different countries.[210]

To conclude, often the activities in these two areas and other advanced broadband applications are still pilot projects. As mentioned before, OECD firms and countries still have a long way to go to seize the full potential of broadband in these areas.

Notes

113 See on this point United States Government Accountability Office (GAO), Telecommunications Report, May 2006, www.gao.gov/new.items/d06426.pdf.

114 See also OECD Science, Technology and Industry Scoreboard 2007 (forthcoming).

115 See the EU Broadband Report, June 2006 at: http://ec.europa.eu/information_society/eeurope/i2010/docs/studies/wp1_report_b roadband_final_draft_june2006.doc.

116 For most European countries, the following industries are included: Manufacturing, Construction, Wholesale and retail, Hotels and restaurants, Transport, storage & communication, Real estate, renting and business activities and Other community, social and personal service activities. For Australia, Agriculture, forestry and fishing, Education and Religious organisations are excluded. For Canada, Agriculture, fishing, hunting and trapping, and Construction - specialist contractors are excluded. For Japan, data refer to enterprises with 100 or more employees and exclude: Agriculture, forestry, fisheries and Mining. Korea includes: Agriculture & Fisheries, Light Industry, Heavy Industry, Petrochemicals, Construction, Distribution, Finance and Insurance, and Other services. For Mexico, data refer to enterprises with 50 or more employees and include: Manufacturing, Services and Construction. For New Zealand, data exclude Government administration and defence, and Personal and other services; the NZ survey also excludes businesses with fewer than 6 employees (calculated by Rolling Mean Employment) and those with turnover of less than NZD 30 000. For Switzerland, data refer to enterprises with 5 or more employees, and include Manufacturing, Construction, Electricity, gas, water, and Services industries. No official data exist for the United States.
Most countries define broadband in terms of technology (*e.g.* ADSL, cable, etc) rather than speed.

117 See the 2006 Yearbook of Information Society Statistics, Republic of Korea, National Information Society Agency, NIA.

118 See 'Internet use and e-commerce in enterprises 2006', Statistics Finland, www.stat.fi/til/icte/2006/icte_2006_2006-10-10_tie_001_en.html.

119 Sensis E-business Report (August 2007): www.about.sensis.com.au/resources/sebr.php) and Sensis, eBusiness Report – The Online Experience of Small and Medium Enterprises, August 2006, www.about.sensis.com.au/resources/sebr.php. as supplied by Australian delegation.

120 Eurostat 2005 Community ICT Household survey at: http://ec.europa.eu/information_society/eeurope/i2010/docs/studies/wp1_report_b roadband_final_draft_june2006.doc.

121　C.f. US Department of Commerce (2004), A Nation Online: Entering the Broadband Age at: www.ntia.doc.gov/reports/anol/NationOnlineBroadband04.htm.

122　NIDA, Korea, December 2006.

123　Statistics Canada, The Daily, 20 April 2007 on 'Electronic commerce and technology', www.statcan.ca/Daily/English/070420/d070420b.htm.

124　OECD Science, Technology and Industry Scoreboard 2007, Paris.

125　See Chapter 3, OECD Information Technology Outlook 2004 and see DSTI/ICCP/IE(2007)3/Rev1 for more details on broadband impacts.

126　See also EU 2007 Annual Reports on i2010 Strategy, http://ec.europa.eu/information_society/eeurope/i2010/docs/annual_report/2007/comm_final_version_sg/com_2007_0146_en.doc.

127　EU report on 'Benchmarking access and use of ICT in European Schools 2006' at http://ec.europa.eu/information_society/eeurope/i2010/docs/studies/final_report_3.pdf

128　In the case of Switzerland figures include only the schools connected by Swisscom.

129　'Public Schools and Classrooms: 1994–2005', U.S. Department of Education, NCES 2007-020, http://nces.ed.gov/pubs2007/2007020.pdf Internet Access in U.S.

130　Statistics Canada, Information and Communications Technologies in Schools Survey, 2003/04 school year, Thursday, June 10, 2004, www.statcan.ca/Daily/English/040610/d040610b.htm.

131　EU report on 'Benchmarking access and use of ICT in European Schools 2006', see fn. 133.

132　Australian National Broadband Strategy Implementation Group, Yearly Update 2006, www.dcita.gov.au/communications_for_business/broadband_and_internet/national_broadband_strategy/NBSIG.
All data from the DCITA, National Broadband Strategy Implementation Group - Yearly Update 2006 (Sourced from MCEETYA ICT in Schools Taskforce, Bandwidth Provision to Australian Government Schools 2002-2006, December 2006).

133　Data as provided by the delegation of the United States of America.

134　NIDA, 2007.

135　Pew Internet and American Life Project study on US wireless Internet access, February 2007, www.pewinternet.org/pdfs/PIP_Wireless.Use.pdf.

136　Some regulators such as ComReg in Ireland now include data on the provision of public and private broadband services over Wi-Fi.

137　See the OECD study on the Participative Web and User-created content, at http://213.253.134.43/oecd/pdfs/browseit/9307031E.pdf.

138　Data obtained from John B. Horrigan, Pew Internet and American Life Project.

139　*Ibid.*

140　See OECD, ICTs and Gender – Evidence from OECD and Non-OECD Countries. UNCTAD Expert Meeting, 4 - 5 December, www.unctad.org/sections/wcmu/docs/c3em29p025_en.pdf.

141　Australian Bureau of Statistics; 8146.0 - Household Use of Information Technology, Australia, 2006-07, 20/12/2007.

142   Data obtained from John B. Horrigan, Pew Internet and American Life Project.

143   Results of the Japanese Telecommunications Usage Trend Survey for 2005, 19 May 2006, The Ministry of Internal Affairs and Communications, www.soumu.go.jp/joho_tsusin/eng/Statistics/pdf/060519_1.pdf.

144   See 'Survey on ICT usage in households and by individuals 2006', Statistics Finland, www.stat.fi/til/sutivi/2006/sutivi_2006_2006-12-11_tie_001_en.html.

145   European Commission, i2010 Annual Report- Germany, http://ec.europa.eu/information_society/eeurope/i2010/docs/annual_report/2007/country_factsheets/2007_factsheet_de.pdf.

146   For example, in the United States, in September 2001, 5.6 percent of rural households had broadband at home; six years later, the rate had reached 38.8 percent, more than a six-fold increase. See National Telecommunications and Information Administration, *Networked Nation: Broadband in America, 2007*, released January 2008, p.14.

147   C.f. OECD work on digital broadband content at www.oecd.org/sti/digitalcontent and European Commission (2006), Interactive content and convergence: Implications for the information society, Study for the European Commission, October 2006, http://ec.europa.eu/information_society/eeurope/i2010/docs/studies/interactive_content_ec2006_final_report.pdf.

148   'Pipe Dreams?: Prospects for next generation broadband deployment in the UK', Report by the UK Broadband Stakeholder Group, 2007 www.broadbanduk.org/content/view/236/7/.

149   C.f. European Commission Staff Working Paper, i2010 Annual Information Society Report 2007Brussels, SEC(2007),Volume 1, http://ec.europa.eu/information_society/eeurope/i2010/docs/annual_report/2007/sec_2007_395_en_documentdetravail_p.pdf.

150   See DSTI/ICCP/IE(2005)3/FINAL and OECD Information Technology Outlook 2006, Paris.

151   OECD (2007) *OECD Communications Outlook 2007*. OECD Publishing, Paris.

152   See the forthcoming OECD study on IPTV, DSTI/ICCP/CISP(2006)5/Rev1 and the OECD Digital Broadband Content: Film and Video (following fn.).

153   See OECD Digital Broadband Content: Film and Video DSTI/ICCP/IE(2006)11/FINAL,Paris. http://213.253.134.43/oecd/pdfs/browseit/9308011E.PDF.

154   C.f. International Telecommunications Union, 'Innovation dynamics in the IP environment', The Future of Voice Conference, January 2007, www.itu.int/osg/spu/ni/voice/papers/FoV-Innovation-Skouby-Tadayoni-Draft.pdf.

155   See chapter 6, OECD Information Technology Outlook 2008 (forthcoming).

156   In the OECD study, 'IPTV: Market Developments and Regulatory Treatment', DSTI/ICCP/CISP(2006)5/REV1, of the OECD Working Party on communication Infrastructures and Services Policy, IPTV is defined as "multimedia services such as television/video/audio/text/graphics/data delivered over IP based networks managed to provide the required level of Quality of Service (QoS)/Quality of Experience (QoE), security, interactivity and reliability." Moreover, it is stated that "VoD service itself does not fall into IPTV services since the service model does not necessarily need a 'required level of QoS/QoE." In practise, ISPs offering IPTV services often include a package of TV channels as well as PayTV and Video on Demand.

[157]   See OECD study, 'IPTV: Market Developments and Regulatory Treatment', DSTI/ICCP/CISP(2006)5/REV1, Paris.

[158]   Ofcom (2006), Communications Market Report 2006, UK Office of Communications, London, www.ofcom.org.uk/research/cm/cm06/.

[159]   'IPTV doubles in the year to Q2 2006', Point Topic report, December 2006, http://point-topic.com/home/press/dslanalysis.asp.

[160]   Ofcom (2007), Communications Market Report 2007, UK Office of Communications, London, 23 August 2007, www.ofcom.org.uk/research/cm/cmr07/cm07_print/.

[161]   "Digital Broadband Content: Mobile Content – New content for new platforms", OECD DSTI/ICCP/IE(2004)14/FINAL.

[162]   See 'Telework Towards Tomorrow: Can Technology Really Change the Workplace?', NMRC MILESTONES Ideas and Trends for the New Millennium, March 2007, http://newmillenniumresearch.org/milestones/march07.html and '2005/2006 National Technology Readiness Survey', Robert H. Smith School of Business' Center for Excellence in Service at the University of Maryland, July 2006, www.smith.umd.edu/ntrs/.

[163]   *Ibid* and 'Telework Trends: Can Technology Really Change the Workplace?', NMRC MILESTONES Ideas and Trends for the New Millennium, Joanne Pratt, Founder & President, Joanne H. Pratt Associates, at www.uwex.edu/disted/conference/Resource_library/proceedings/05_1864.pdf.

[164]   Provided by the Canadian Delegation, based on the Canadian Telework Association www.ivc.ca/cta/index.htm.

[165]   Sensis Insights Teleworking Report (business and consumer survey results) and Sensis Business Index, Sweeney Research - May 2006, prepared for DCITA (Australian Ministry), www.dcita.gov.au/communications_for_consumers/internet/telework/australian_telework_advisory_committee/sensis_insights_report_teleworking.

[166]   Eurostat.

[167]   While *e.g.* many companies and administrations make an effort on site to curb energy consumed by IT, this is not necessarily the case when a person works from home where the energy consumption increases.

[168]   Eurostat 2005 Community ICT Household survey.

[169]   OECD (2005), e-Government for Better Government, www.oecd.org/document/45/0,2340,en_2649_201185_35815981_1_1_1_1,00.html.

[170]   C.f. Capgemini (2006), 'Online Availability of Public Services: How Is Europe Progressing?', Web Based Survey on Electronic Public Services, Report of the 6th Measurement prepared by Capgemini for the European Commission, June 2006, Belgium and Accenture (2006), 'High Performance in Government', Report, www.accenture.com/NR/rdonlyres/D7206199-C3D4-4CB4-A7D8-846C94287890/0/gove_egov_value.pdf.

[171]   Japan White Paper, 2006.

[172]   OECD (2005), *e-Government for Better Government*.(OECD, Paris), Annex 4.A1 and OECD (2006), e-Government Studies Denmark (OECD, Paris).

[173]   Accenture (2006), 'High Performance in Government', Report, www.accenture.com/NR/rdonlyres/D7206199-C3D4-4CB4-A7D8-846C94287890/0/gove_egov_value.pdf. Future OECD work will focus on the take-up of e-government services, see the OECD Programme of Work and Budget

2007 – 2008 of the Public Governance Committee. See also "Online Availability of Public Services: How is Europe Progressing?", European Commission, June 2006,
http:/europa.eu.int/information_society/eeurope/i2010/docs/benchmarking/online_availability_2006.pdf.

[174] OECD (2004), *OECD e-Government Studies Finland* (OECD, Paris); OECD (2005), *OECD e-Government Studies Norway* (OECD, Paris); OECD (2005), *OECD e-Government Studies Mexico* (OECD, Paris); OECD (2006), *OECD e-Government Studies Denmark* (OECD, Paris); forthcoming OECD (2007), *OECD e-Government Studies Hungary* (OECD, Paris); forthcoming OECD (2007), *OECD e-Government Studies The Netherlands* (OECD, Paris); forthcoming OECD (2007), *OECD e-Government Studies Turkey* (OECD, Paris).

[175] Accenture (2006), 'High Performance in Government', Report,
www.accenture.com/NR/rdonlyres/D7206199-C3D4-4CB4-A7D8-846C94287890/0/gove_egov_value.pdf.

[176] Danish Ministry of Science, Technology and Innovation, "Digital Communication between Citizens and the Public Sector", 2006.

[177] See, for instance, e-learning Nordic 2006
http://itforpedagoger.skolutveckling.se/in_english; Eight Years of ICT in Schools, Ministry of Education Culture and Science, Netherlands.

[178] Eurostat data. In the EU15, 36% of Internet users in broadband households and 30% of Internet users in non-broadband households use Internet for education.

[179] Data from a recent US survey among school superintendents and school technology directors suggest that the bandwidth in schools should increase from the current 2,9 kilobits per second per pupil to more than 9 in five years, although some analysts believe that this width should increase up to 40 kilobits per second per pupil. America's Digital Schools 2006. A five year forecast. Mobilizing the curriculum. The Greaves Group. The Hayes Connection, Littleton, CO, www.ads2006.org/main/pdf/ADS2006KF.pdf.

[180] See, for example, the "Broadband Blueprint—Broadband Development", Department of Communications, Information Technology and the Arts – Australia, 2006: at:
www.dcita.gov.au/communications_for_consumers/internet/broadband_blueprint/broadband_blueprint.

[181] See "Broadband Blueprint—Broadband Development", Department of Communications, Information Technology and the Arts – Australia, 2006: at:
www.dcita.gov.au/communications_for_consumers/internet/broadband_blueprint/broadband_blueprint.

[182] See, *e.g.* Australian study 'Broadband in Health: Drivers, Impediments and Benefits', at
www.dcita.gov.au/communications_and_technology/publications_and_reports/2002/august/broadband_in_health_drivers,_impediments_and_benefits Standardisation and Interoperability.
The potential impact of broadband on the economics of health care was explored is a study by Robert Litan. This author estimates that over the next 25 years, broadband-based health applications could result in savings of at least USD 927 billion in the health care costs for seniors and the disabled. See also Robert E. Litan (2005), 'Great Expectations: Potential Economic Benefits to the Nation

from Accelerated Broadband Deployment to Older Americans and Americans with Disabilities', New Millennium Research Council, December 2005, www.newmillenniumresearch.org/archive/Litan_FINAL_120805.pdf, Robert E. Litan (2006), 'Massive Economic Benefits Foreseen: Ultra-fast telemedicine and telecommuting can save money and improve quality of life' at www.broadbandproperties.com/2006issues/feb06issues/Litan %20 %20 Health %20and %20Medicine.pdf, February 2006. See also studies at www.newmillenniumresearch.org/news/Litan_Hill_PR040406.pdf and wwww.newmillenniumresearch.org/archive/Telemedicine_Report_022607.pdf.

183   Robert E. Litan (2006), 'Massive Economic Benefits Foreseen: Ultra-fast telemedicine and telecommuting can save money and improve quality of life' at: www.broadbandproperties.com/2006issues/feb06issues/Litan %20 %20Health %20and %20Medicine.pdf, February 2006.

184   Good Health Services across Europe – Evidence on their economic benefits and lessons learned, Tom Jones, Alexander Dobrev, Karl A. Stroetmann, Report for the European Commission, www.ehealthconference2006.org/pdf/good_eh_proc.pdf.

185   See Chapter 5, OECD Information Technology Outlook 2004. See also Report on *'Finding Answers Online in Sickness and in Health'*, Pew Internet and American Life Project report, at www.pewInternet.org/pdfs/PIP_Health_Decisions_2006.pdf. In relation to health, Pew reported that 26 % of adult Internet users who had dealt with the issue in the previous two years said that the Internet played a crucial or important role as they helped another person cope with a major illness; the equivalent figure for themselves as they coped with a major illness was 28%. The December 2005 Pew survey found that one in five adult Internet users reported that the Internet had greatly improved the way they get information about health care, with a variety of Web-based information sources used.

186   See www.health.gov.au/Internet/wcms/publishing.nsf/Content/health-publicat.htm.

187   'Barriers and drivers of on line medical applications: A view from the health industry', Rinde & Balteskard (2002), www.ptc.org/PTC2004/program/private/monday/m13/m133_firth.pdf.

188   C.f. OECD (2003), *Digital delivery of goods and services in healthcare*, DSTI/ICCP/IE(2002)13/FINAL and the OECD Information Technology Outlook 2004, Chapter 5, www.oecd.org/dataoecd/22/18/37620123.pdf.

189   "Broadband for growth, innovation and competitiveness" The IT Policy Strategy Group – Sweden, at: www.sweden.gov.se/sb/d/574/a/76048;jsessionid=aRgt9J6DAf-g.

190   "Information Society Action Plan 2006-2010", State Planning Organisation – Turkey, July 2006 at: www.dpt.gov.tr/konj/DPT_Tanitim/pdf/Information_Society_Strategy.pdf.
   Additional information is available at www.connectednation.com/index.php and http://connectkentucky.org/.

192   www.nederlandbreedbandland.nl/

193   *Source*: UK Department for Business, Enterprise and Regulatory Reform and Ofcom, available at: www.berr.gov.uk/files/file37744.pdf.

194   See Project PROBE Case Study, A case Study of Project PROBE, delivering broadband to rural schools, www.e.govt.nz/resources/research/case-studies/project-probe.

195   See www.universalservice.org/sl/.

196   The government of Greece also plans to provide for broadband interconnection between schools, but also to culture and sports centers (including municipal libraries, museums), public bodies/ agencies and regional surgeries, health centers, and so on. Less populated municipalities are to be provided with wireless broadband access. See Greek broadband plan.

197   EU document. Brussels, 12.5.2004COM(2004) 369 final communication from the commissionto the council, the european parliament,the european economic and social committee and the committee of the regions. Connecting Europe at High Speed: National Broadband Strategies.

198   COM(2006) 129 final at: http://eur-lex.europa.eu/LexUriServ/site/en/com/2006/com2006_0129en01.pdf.

199   Bakkom, Stratégie du Conseil fédéral pour une société de l'information en Suisse, janvier 2006, www.bakom.ch/themen/infosociety/00695/index.html?lang=fr.

200   "Information Society Action Plan 2006-2010", State Planning Organisation – Turkey, July 2006 at: www.dpt.gov.tr/konj/DPT_Tanitim/pdf/Information_Society_Strategy.pdf.

201   See www.hoykom.no/.

202   See www.dti.gov.uk/sectors/digitalcon/effectivecontent/page10175.html.

203   Policies evolve around the following areas: innovation and technology; value chains and business models; enhancing the infrastructure; the business and regulatory environment; supply and use of public sector information and content; and conceptualisation, classification and measurement. See OECD (2006), "Digital Broadband Content: Digital Content Strategies and Policies", DSTI/ICCP/IE(2005)3/FINAL, http://ww.oecd.org/dataoecd/54/36/36854975.pdf.

204   *Ibid*. For a full list of policies see: www.oecd.org/countrylist/0,3349,en_2649_34223_38711225_1_1_1_1,00.html.

205   See www.dcita.gov.au/communications_for_business/broadband_and_internet/interoperability_-_building_the_case_for_e-business.

206   See www.etri.re.kr/www_05/org/e_main.htm?pagecode=0202&url=../org/e_dhrd/team/team_02.html. and www.etri.re.kr/www_05/org/e_main.htm?pagecode=0101&url=../org/e_trd/intro/intro_01.html

207   See http://198.103.246.211/demoprojects/demo_aboriginal_e.asp and www.health.gov.au/internet/wcms/publishing.nsf/Content/health-ehealth-broadband-index.htm.

208   See https://projects.ibbt.be/coplintho/index.php?id=131.

209   See "HØYKOM Introduction", Available at: www.hoykom.no/hoykom/hoykomweb.nsf/4a87ff3bf2c03cc38525646f0072ffa9/42de84fc9e2014aac125700c00497969/$FILE/Evaluation%20_e.pdf.

210   ICT as a mode of transport A review of the use of information and communication technology to achieve transport policy goals, British Telecom Report, Forum for the Future, www.btplc.com/Societyandenvironment/Reports/ICTasamodeoftransportfinal.pdf.

Chapter 3

THE FRAMEWORK ENVIRONMENT FOR BROADBAND

Security, privacy and consumer protection

Market trends and developments since 2003

There have been a number of security developments emerging in OECD markets over the last three years. First, spam has evolved from a simple nuisance to a vehicle for fraud. It is has become the primary vector for introducing "malware" – malicious software into computers. Second, once a computer is infected with malware it can be remotely controlled to launch cyber attacks on defined targets – unbeknownst to the user.[211] Clusters of these infected computers are called "botnets" and can be controlled simultaneously to launch attacks which overwhelm targeted Internet servers until they cease to function.

The broadband transition to fibre connections and symmetric bandwidth could make botnet attacks much more virulent. One infected computer on a fibre connection with 100 Mbit/s of upload capacity could theoretically cause as much damage as 390 infected computers with upload speeds of 256 kbit/s. The average advertised upload speeds for broadband in the OECD in October 2006 was 1 Mbit/s for DSL, 0.7 Mbit/s for cable and 31 Mbit/s for FTTx. This implies that the potency of one infected computer on a fibre connection would be equivalent to 31 infected computers on DSL and 44 computers on cable networks. Clearly, the move to faster upload bandwidth via fibre could make the botnet problem much more severe. Most consumers lack the level of sophistication or training to understand the magnitude of the threats and the various tools available to them to protect against them. Under certain circumstances, network providers can help prevent consumers from being vulnerable to an increasing array of sophisticated forms of computer malware.

There also have been a number of privacy developments in the last three years. Broadband's always-on connections mean that consumers spend more time online making more financial and commercial transactions over the Internet. Moreover, the Internet, and the web in particular, have also become more interactive and more personal information about users is stored online. One evolution has been the emergence of social networks (see the section on usage). Through these services, individuals share personal information including multimedia and content with other individuals or the broader public. Users typically have some control over how their information is displayed to other users but the manner in which companies treat users' personal data varies.[212] There has been some concern raised about whether users fully understand the contracts they agree to when they use content-distribution sites and if they are aware of how their personal information and content may be used in the future.[213] In addition, thieves have targeted online retailers in an effort to steal personal and credit card information from shoppers. A number of significant data breaches have taken place which has affected millions of customers.[214]

As raised earlier, consumer protection issues related to broadband are coming to the fore. Broadband contracts may have high penalties if consumers leave or service could be disconnected for long periods when consumers switch providers. Consumers find it difficult to switch between broadband suppliers or to move residences without experiencing problems. Some consumers reportedly lost their broadband service for several weeks or received confusing and contradictory information about what they needed to do in order to migrate to another ISP.[215]

Application of the Recommendation

The OECD Council Recommendation calls for a culture of security to: enhance trust in the use of ICT by business and consumers; improve enforcement of privacy and consumer protection; and more generally, strengthen cross-border co-operation between stakeholders to reach these goals. OECD governments and industry have developed more initiatives to promote a culture of security. Technological solutions (content filters for minors, spam filtering, anti-virus protection, etc.) have also come to the fore.

Security of information systems and networks

Ensuring the security of information systems and networks is an important area of focus for most OECD governments. Information security policies in many OECD countries entail various elements (see Table 3.1): threat monitoring mechanisms; advisory bodies; computer emergency response teams; anti-spam and anti-phishing initiatives; verification mechanisms; ICT security standards; a public key infrastructure: cybercrime legislation and security-related R&D.

Table 3.1. Broadband and network security policy examples from OECD countries

1. Monitoring threats: the monitoring of security threats, available technical solutions and how business, citizens or the government are prepared to handle such threats[216]	**2. Advisory bodies:** the creation of co-ordinating and advisory bodies formed at the highest levels of government on ICT security and for the protection of critical information infrastructures[217]	**3. CERTs:** the establishment of special centres to detect incidents and constituting computer emergency response teams[218]
4. Anti-spam, -phishing: initiatives against spam and phishing[219]	**5. Verification mechanisms:** certification mechanisms whether in the form of authentication for businesses, electronic certification platforms for businesses and citizens[220], or the establishment of protocols for digital signatures[221] and associated legal frameworks[222]	**6. ICT security standards:** standards for ICT security processes in the government sector
7. Protecting PKI: the protection of public key infrastructure (PKI)	**8. Cybercrime legislation:** the development of legislation dealing with cybercrime [223], ICT security breaches (o.g. legislation with provisions against hacking, computer viruses, data interference) and providing police officers with the required skills[224]	**9. Security-related R&D:** support schemes for ICT security R&D and fostering the development of the ICT security industry[225]

Source: National broadband policies and other ICT policies.

Better policies and frameworks (especially those co-ordinated at the international level), increased law enforcement, improved responses by Computer Security Incident Response Teams and proactive measures will be needed in the future.[226] OECD countries will have to intensify their efforts.

Privacy

Privacy protection is also recognised as a priority, with several different patterns of activity. The development of new legislation on the privacy of individuals and the transfer and sharing of personal data (*e.g.* the *Korean Personal Information Protection Act*[227], or the implementation of European Union directives in this area) has been pursued by some OECD governments. Other notable actions include increasing cross-border enforcement co-operation and the development of policy measures, like educational initiatives and privacy-enhancing technologies.

Protecting consumers

Consumer protection efforts of OECD governments are also concentrated in select areas. The main goal of these initiatives is to promote confidence and to encourage wide broadband take-up.[228] Foremost is the development of awareness campaigns to educate consumers about risks and measures they can take to protect against fraudulent practices and increase their security on the Internet, *e.g.* the *National Information Security Day* in Finland,[229] and the Australian *National E-Security Awareness.*[230]

OECD countries also pursue the establishment of legal frameworks which specifically address questions of e-commerce, while others are derivations or expansions of existing laws on consumer protection (*e.g.* the protection of children). OECD countries have also recently agreed to a new approach to make online shopping safer (OECD Recommendation on Consumer Dispute Resolution and Redress).[231]

Policy makers are increasingly interested in examining new consumer protection issues related to broadband.[232] OECD governments aim to intensify competition. They are also concerned with providing consumers transparency, the right information to compare and choose broadband providers, and with offering them redress in the case of problems with broadband providers.

The National IT and Telecom Agency in Denmark issues a quarterly price report with the aim of generating better price comparisons in the telecommunications market.[233] Both the Australian and United Kingdom telecommunications industries have "ombuds" offices for consumer complaints. The Advertising Standards Authority in the United Kingdom asked the ISPs to indicate in advertisements that connection speeds could vary considerably. In the United States, the Federal Communications Commission partly carries out this function. There are recommendations in Canada to establish an independent "ombuds" office that should have authority to resolve complaints from individual and small business retail customers of any telecommunications service provider.[234] In Finland, a working group was appointed to survey consumer policy problems in information society services, with the goal of offering consumers compensation for errors and delays in the telecommunications service and triggering regulatory amendments.[235] The legislation in some OECD countries now requires providers to offer independent alternative dispute resolution (ADR) schemes.

A few OECD governments have also looked at the various lock-in periods for broadband subscriptions. The Swedish National Post and Telecom

Agency, for instance, is already working actively on the production of comparative quality information for consumers and on preventing prohibitive contractual clauses.[236] In Denmark, operators are not permitted to utilise lock-in periods for private end users for periods exceeding six months. Another method of doing this is to encourage the labeling of broadband products and services with various quality declarations and quality standards.

Consumers may also suffer from restricted or downgraded access to particular types of content or applications if broadband service providers begin discriminatory traffic shaping or blocking certain services on their networks. Several OECD governments are contemplating measures to ensure that consumers are entitled to run applications and services of their choice.[237] There are recommendations in Canada that regulatory frameworks should include provisions that confirm and protect the right of Canadian consumers to access publicly available Internet applications and content of their choice by means of public telecommunications networks that provide access to the Internet.[238]

OECD studies have raised various consumer issues (*e.g.* interoperability problems) with respect to broadband content.[239] Counteracting illegal and harmful content on the Internet and protecting children online have also been recent priorities for many OECD governments. Information and attitude-building projects (for example, the SAFT-project – safety and awareness in Norway) are contributing to this goal. A number of countries have also focussed on providing age rating systems, including government or industry self-regulatory initiatives to develop national and transnational rating systems. Several campaigns are ongoing to help parents monitor the Internet activities of their children (*e.g.* the German *"SCHAU HIN! Was Deine Kinder machen"* – campaign). OECD governments have also encouraged codes of conduct to guide self-regulatory approaches and cross-border initiatives in some cases.

Regulatory frameworks balancing the interests of suppliers and users

Market trends and developments since 2003

The Recommendation calls for regulatory frameworks that balance the interests of suppliers and users in areas such as the protection of intellectual property rights (IPRs) and digital rights management (DRM), without disadvantaging innovative e-business models.

Application of the Recommendation

In recent years, this policy objective has grown in importance and will remain a top priority – even after other policy goals, such as achieving widespread broadband access, have been achieved.

New types of IPR issues are appearing. As old barriers between physical distribution and broadcasting and Internet distribution crumble, new technologies and new ways of creating, accessing and distributing content are emerging, and new enforcement challenges are thus materialising. This necessarily creates tensions with the rise of new technologies and new business models, but also new possibilities for infringing uses.

With these shifts a wide range of new industry and policy issues have arisen. For example, while artists and the communications industry are struggling to prevent digital piracy, DRM technologies that can protect content and enable new business models have run into practical difficulties and consumer resistance.[240] New IPR-related challenges are emerging every day. Examples are the complexities in creating international digital libraries; search engines being sued for archiving content from newspapers; the liability of online intermediaries when it comes to pirated material or harmful content; the boundaries of fair use, fair dealing and other exceptions and other limitations to copyright (especially in areas such as user-created content or education); the application of copying levies on new electronic devices or Internet services; and the rise of alternative licensing schemes.

OECD governments have been active in trying to strike a balance between setting the right incentives for creation and diffusion of protected works. Governments have worked to promote the protection of IPRs through legislation (national law and international treaties such as the ratification of the WIPO Internet Treaties), enforcement, increased criminal sanctions and awareness/education campaigns (including youth education programmes). International efforts to protect IPR continue to play an important role.

Moreover, to achieve the right balance between creation and diffusion and use, OECD governments have been monitoring and analysing new developments in the area of intellectual property (*e.g.* the UK Gowers Review of Intellectual Property)[241] and, for example, holding public consultations. In areas such as DRM, parliamentary and government study groups have analysed existing opportunities and challenges for rights-holders and users (*e.g.* the EU High-Level Group on DRM).[242] When deemed appropriate, OECD governments have created new exemptions to the prohibition on circumvention of copyright protection systems for access control technologies, *e.g.* in the United States for educational use. New and facilitated ways of licensing and new technologies are being explored.

OECD governments are also seeking solutions to put more government or public content online. For instance, as part of the European Digital Library initiative an EU expert group on digital libraries has agreed to a basic model for handling copyrights for digitalised cultural publications in libraries. The government of Australia is looking into whether the current IPR regime (*i.e.* the Crown copyright regime) is too restrictive preventing government content from going online.

OECD governments also facilitate dialogue and consensus-finding among different industry participants in the value chain who are involved, for example, in developing new online business models and reducing online piracy (*e.g.* the Italian San Remo charter, the French Rapport Olivennes, the EU Communication on Creative Content Online).[243]

Many of the above issues will play out in the market place without government involvement. Yet OECD governments are well advised to continue closely monitoring developments and adjusting the regulatory system when necessary. The new technological environment and new forms of content creation and diffusion may best be complemented – in certain cases – by innovative policies that call into question existing approaches in the field of IPR. It is crucial that economic analysis underpins the proposed regulatory modifications in IPR regimes. It is also important that processes be open and that content creators and consumers are full stakeholders in this process.

Research and development (R&D)

Market trends and developments since 2003

The Recommendation encourages R&D in the field of ICT for the development of broadband and enhancement of its economic, social and cultural effectiveness.

The ICT sector is very R&D-intensive. This is reflected by the fact that ICT manufacturing industries account for roughly more than a quarter of total manufacturing business R&D expenditure in most OECD countries, more than half in Finland, and Korea, and more than 30% in the United States, Canada and Ireland. However, business R&D expenditures still vary widely between OECD countries (see Figure 3.1).[244] More importantly, overall business investment in R&D still accounted for a small share of GDP (less than 1.3% in manufacturing sector and less than 0.2% in the ICT services sector). Hence, there is room for further improvement on meeting the policy goal of the OECD Recommendation.

Figure 3.1. Business R&D expenditure, 1997 and 2004

Business R&D expenditure by selected ICT manufacturing industries
As a percentage of GDP

◆ 1997 ☐ 2004

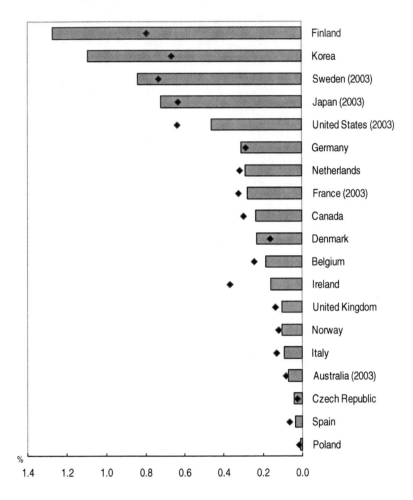

Note: Owing to unavailability of R&D for class 642 (Telecommunications), division 64 (Post and telecommunications) is used as a proxy. Available information shows that in the United States, class 642 accounts for 97-98% of the R&D in division 64.

Figure 3.1. Business R&D expenditure, 1997 and 2004 *(continued)*

Business R&D expenditure by selected ICT services industries

As a percentage of GDP

◆ 1997   ▢ 2004

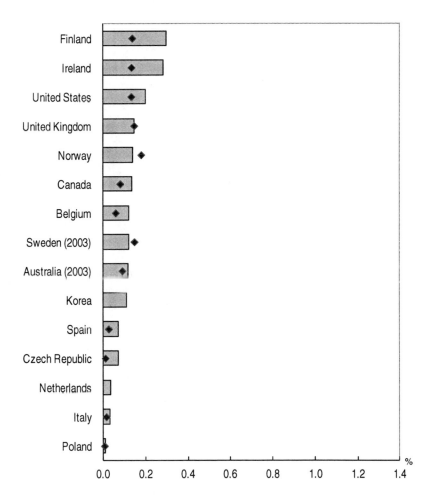

Note: Owing to unavailability of R&D for class 642 (Telecommunications), division 64 (Post and telecommunications) is used as a proxy. Available information shows that in the United States, class 642 accounts for 97-98% of the R&D in division 64.

Application of the Recommendation

In general, most if not all OECD countries emphasise that ICT-related R&D is a very high and an increasing important policy priority. OECD member country support is generally focussed on:

* R&D in the ICT sector of which some is relevant to broadband;

* ICT R&D in the context of applications to other non-ICT sectors (*e.g.* embedded systems in the automobile sector); and

* Use of broadband to conduct research (*e.g.* e-science, research networks).

Support of broadband networks and applications

OECD countries emphasise research and innovation in the fields of broadband infrastructure (*e.g.* networks, connecting technologies, system support products and testing), related applications (especially in the wireless area), broadband-enabled public services, digital broadband content and even R&D focusing on new broadband business models.

Special research groups have emerged in these fields. The United States National Science Foundation's' Global Environment for Network Innovations (GENI) is designed to allow experiments on a wide variety of problems in communications, networking, distributed systems, cyber-security, and networked services and applications.[245] In Belgium, the Interdisciplinary Institute for Broadband Technology is a research institute focusing on broadband technology.

Increasingly OECD governments are also encouraging R&D and innovation in broadband content. Through the DTI's Technology Programme, the UK government is providing "smart funds" to encourage innovation and research in developing broadband content (see also the Austrian Wireline R&D Programme).[246] In Korea, digital content, embedded software, and telematics are some of the three pillars in its new ICT growth strategy (IT389). The Finnish National Technology Agency allocated EUR 39 million for mobile entertainment projects, technology development and game production. When not exclusively related to broadband, many OECD governments have developed ICT R&D projects with direct impacts and links to broadband. As part of its High-Tech Strategy, Germany supports the Internet of Things, online intelligence search and management of digital content, IT security, virtual reality, grid computing, smart energy management, and mobile applications in public administration.[247]

Research networks, grids and e-science

Broadband is also recognised as necessary tool for research. More and more OECD countries have programmes for advanced research networks and are developing high-capacity research backbones. At the European level, shared infrastructures such as GÉANT are in place. The Dutch GigaPort project, facilitated by the government, provides one of the fastest R&D networks in the world.[248] The Japan Gigabit Network (JGN II) is the advanced test-bed network for R&D and development.[249] As a public-private partnership, the Canadian Network for the Advancement of Research, Industry and Education (CANARIE) helps raise and allocate funds for research network and Internet development in Canada.[250]

Also, governments and public institutions are exploring new ways of digital access to scientific information and are providing access to research.

Existing ICT-related R&D initiatives face the following challenges. First, one may wonder whether the overall level of funding in this area is sufficient considering the economic and social importance of broadband. The role of government in basic R&D may have to be reaffirmed and falling government contributions in this area should be assessed.[251] While governments often pledge to increase R&D, actual spending levels often fall short of policy goals. In addition, the question is whether broadband receives the necessary policy priority. Second, the adequacy of existing government R&D support schemes (*e.g.* R&D tax credits) may be questioned, as we move more towards research related to networks, broadband services and content. Third, the strengthening of research networks, their applied use, and improving scope for international co-operation in this field should be a policy priority. Fourth, the co-ordination and alignment of federal government, sub-national, and private R&D programmes sometimes prove challenging.

Notes

[211] "Analytical Report on Malware", OECD DSTI/ICCP/REG(2007)5, OECD, Paris.

[212] See the OECD study on the Participative Web and User-created content, at http://213.253.134.43/oecd/pdfs/browseit/9307031E.pdf.

[213] *Ibid.* and "The Content in Google Apps Belongs to Google", Joshua Greenbaum – ZDNet, 28 August 2007, at: http://blogs.zdnet.com/Greenbaum/?p=130.

[214] "Credit card breach exposes 40 million accounts", ClNet News, 17 June 2005 at: http://news.com.com/Credit+card+breach+exposes+40+million+accounts/2100-1029_3-5751886.html and "45.7m card details stolen in TJX security breach", Computer Weekly, 30 March 2007, at: www.computerweekly.com/Articles/2007/03/30/222778/45.7m-card-details-stolen-in-tjx-security-breach.htm.

[215] "Broadband Migrations: Enabling Consumer Choice", Ofcom – United Kingdom, 13 December 2006, at: http:// ofcom.org.uk/consult/condocs/migration/statement/statement.pdf.

[216] "La pénétration des technologies de l'information au Luxembourg", Mindforest, at: http://eco.public.lu/documentation/publications/reperes/Reperes.pdf.

[217] *E.g.* in Australia and Germany. Swiss examples for monitoring threats would be www.melani.admin.ch and www.cybercrime.ch.

[218] *E.g.* the 'US CERT'.

[219] *E.g.* in Japan , or the Danish strategy against spam) – including on the international level such as the OECD Anti-Spam Toolkit.

[220] *E.g.* LuxTrust in Luxembourg.

[221] "IT and Telecommunications Policy Report 2007", Danish Government, April 2007, at: http://videnskabsministeriet.dk/site/forside/publikationer/2007/it-and-telecommunications-policy-report-2007/IT %20and %20Telecommunications %20Policy %20Report %202007.pdf.

[222] *E.g.* the Korean Electronic Signature Act. See "2006 Korea Internet White Paper", Ministry of Information and Communication – Korea, May 2006, at: http://eng.mic.go.kr/eng/secureDN.tdf?seq=10&idx=1&board_id=E_04_03.

[223] *E.g.* the European Convention on Cybercrime.

[224] *E.g.* in France at www.internet.gouv.fr/information/information/actualites/5e-comite-interministeriel-pour-societe-information-296.html.

[225] *E.g.* the German 2010 strategy.

[226] OECD (forthcoming), Paper on Malware, DSTI/ICCP/REG(2007)5/REV2, Paris.

[227] "2006 Korea Internet White Paper", Ministry of Information and Communication – Korea, May 2006, at: http://eng.mic.go.kr/eng/secureDN.tdf?seq=10&idx=1&board_id=E_04_03.

[228] "Stratégie du Conseil fédéral pour une société de l'information en Suisse", BACOM, January 2006.

[229] "National broadband strategy: Final report", Ministry of Transport and Communications Finland, 23 January 2007 at: http://mintc.fi/oliver/upl615-LVM11_2007.pdf.

[230] LU-Official-2-Reperes_2006.pdf[230] - HOTLINE FOR YOUNG PEOPLE ABOUT INFO SECURITY"mySecureIT", which aims to increase awareness among young people about information security. A "HOTLINE" was set up to ask questions or report any incidents.

[231] See http:// oecd.org/dataoecd/43/50/38960101.pdf for the Recommendation.

[232] "Enhancing competition in telecommunications: Protecting and empowering consumers", OECD DSTI/ICCP/CISP(2007)1, Paris.

[233] The National IT and Telecom Agency also provides a price guide on the Internet, www.teleprisguide.dk.

[234] "Telecommunications Policy Review Panel Final Report 2006", Industry Canada, March 2006,
www.telecomreview.ca/epic/site/tprp-gecrt.nsf/vwapj/00A_e.pdf/$FILE/00A_e.pdf.

[235] "National broadband strategy: Final report", Ministry of Transport and Communications Finland, 23 January 2007, www.mintc.fi/oliver/upl615-LVM11_2007.pdf.

[236] "Broadband for growth, innovation and competitiveness" The IT Policy Strategy Group – Sweden, www.sweden.gov.se/sb/d/574/a/76048;jsessionid=aRgt9J6DAf-g.

[237] Federal Communications Commission (FCC), policy statement FCC 05-151, adopted 5 August 2005.

[238] "Telecommunications Policy Review Panel Final Report 2006", Industry Canada, March 2006:
www.telecomreview.ca/epic/site/tprp-gccrt.nsf/vwapj/00A_e.pdf/$FILE/00A_e.pdf.

[239] See studies at www.oecd.org/sti/digitalcontent.

[240] For disclosure issues see OECD (2006), Report On Disclosure Issues Related To The Use of Copy Control And Digital Rights Management Technologies, OECD Consumer Policy Committee, www.oecd.org/dataoecd/47/31/36546422.pdf.

[241] UK Treasury (2006), Gowers Review of Intellectual Property, Counsellor of the Exchequer, United Kingdom, December 2006:
www.hm-treasury.gov.uk/media/6/E/pbr06_gowers_report_755.pdf, December 2006.

[242] The EC also set up the INDICARE project (The Informed Dialogue about Consumer Acceptability of Digital Rights Management Solutions) to address the consumer side of managing rights.

[243]. See Italian efforts at
www.innovazione.gov.it/ita/news/2005/cartellastampa/sanremo/index.shtml or the French Rapport Olivennes at:
www.culture.gouv.fr/culture/actualites/index-olivennes231107.htm. See also Commission of the European Communities, Communication on Creative Content Online in the Single Market, 3 January 2008, SEC(2007) 1710 at:
http://ec.europa.eu/avpolicy/other_actions/content_online/index_en.htm.

[244] See forthcoming OECD Information Technology Outlook 2008, chapter 3.

245 Information on the GENI project is available at: www.geni.net/.

246 Connecting the UK: the Digital Strategy, Prime Minister's Strategy Unit, March 2005, A joint report with the Department of Trade and Industry, www.cabinetoffice.gov.uk/strategy/work_areas/digital_strategy/index.asp.

247 Information Society Germany 2010 (iD2010), ICT 2020 and German High Tech Strategy, www.bmbf.de/pub/bmbf_hts_lang_eng.pdf.

248 "Broadband and Grids Technology in the Netherlands", Ministry of Economic Affairs, 2005 at: www.hightechconnections.org/2005/broadband.pdf.

249 See also "The Next-Generation Broadband Strategy 2010", Ministry of Internal Affairs and Communications - Japan, August 2006.

250 See www.canarie.ca/about/about.html for more information.

251 See *e.g.* National Academies (2003) Innovation in Information Technology, Computer Science and Telecommunications Board (CSTB), Computer Science and Telecommunications Board (2007) Toward a Safer and More Secure Cyberspace, Seymour E. Goodman and Herbert S. Lin, Editors, National Research Council.

Chapter 4

BROADBAND POLICY ASSESSMENT AND EVALUATION

Need for improved policy assessment and evaluation

Evaluating procedures in formulating ICT policy goals and instruments is increasingly important in OECD countries; and evaluation has become a more common policy tool.

Over the previous three years, some OECD countries have set up annual reviews or study groups to periodically reassess regulations and market developments. OECD countries are convening expert groups to review the market and make policy suggestions. For example, in the UK the government set up its Broadband Study Group (BSG) as an independent advisory body on the development and implementation of a suitable broadband strategy. Other countries have formed similar groups to propose ideas and strategies for promoting broadband development without excessive government influence. Examples of other countries with broadband study groups, or commissions which deal with broadband, include Belgium,[252] Japan, Canada,[253] the Czech Republic,[254] Finland,[255] France, Ireland, Italy,[256] the Netherlands,[257] and Sweden.[258] The European Commission has long encouraged EU Member States to adopt and implement national broadband strategies as a way to "stimulate the supply and the demand sides of the market whenever identified as a national priority".[259]

However, only a few countries have specific broadband policy assessment and evaluation activities which mean that broadband plans can be implemented in a more effective and accountable manner. Plans rarely include mechanisms to review the performance of government initiatives.

There are exceptions. In 2005, the Finnish government adopted a new resolution specifying the objectives of the national strategy. This allowed the group to monitor the implementation of this national strategy and to produce a public report detailing its findings,[260] Japan[261] and Korea[262] also make evaluation of their strategies a key element of their national plans. Canada's Telecommunication Policy Review Panel highlights how formal evaluation requirements should be built into government programmes and how efforts should be made to learn lessons from previous connectivity programmes, especially those that are still in place.[263] The Swiss,[264] Italian,[265] Turkish,[266] Belgian[267] and Swedish[268] governments have also stressed the importance of evaluation.

Statistics and analysis

It is important to develop appropriate and internationally comparable broadband metrics. Policy makers need a wider range of quality indicators to see where their own policies are succeeding and where adjustments may be necessary. Without them, it is difficult for policy makers to judge the level of development in their own domestic markets and to observe trends emerging globally. Ofcom and the Broadband Study Group in the UK have come to a consensus on six indicators that the government should provide on broadband.

The current criteria that the OECD uses to measure broadband (download speeds of at least 256 kbit/s) will need to be reconsidered and possibly changed, as new applications appear which require faster connections than the lowest-speed broadband connections can support. One key challenge will be to devising a new functional definition, particularly as telecommunication markets converge and next-generation networks evolve. Policy makers will need to assess whether it will be necessary to move towards criteria based on possible applications rather than a certain bit rate.[269]

Improving statistics on the uptake and the usage of broadband applications and services is also important. Data and analysis on obstacles to uptake are of increasing interest. The government of Finland, for example, recently assessed the legal and administrative obstacles to the spreading of new broadband technologies and services.[270] The French Broadband Observatory is studying the behaviour of content users/creators.[271]

Need for improved policy co-ordination and sharing of best practises among OECD countries

Improved policy co-ordination among various agencies, ministries and the private sector will be essential. This holds especially true for advanced broadband applications in vital sectors such as health, transport and others where responsibilities are shared. It is also important that policy makers also look beyond their national borders to find best practices in other member countries.

Notes

252 Opinion of the Consultative Committee on Telecommunications of 31/01/2007 "How to improve broadband penetration in Belgium?", at: Dutch version: www.rct cct.bc/docs/adviezen/NL/070131%20breedband%20nl.pdf; French version: www.rct-cct.be/docs/adviezen/FR/070131%20breedband%20fr.pdf.

253 "Telecommunications Policy Review Panel Final Report 2006", Industry Canada, March 2006, at: www.telecomreview.ca/epic/site/tprp-gecrt.nsf/vwapj/00A_e.pdf/$FILE/00A_e.pdf.

254 "The National Broadband Access Policy", Czech Republic Ministry of Informatics, 2005, at: www.micr.cz/files/2185/MICR_brozura_en.pdf.

255 "National broadband strategy: Final report", Ministry of Transport and Communications Finland, 23 January 2007 at: www.mintc.fi/oliver/upl615-LVM11_2007.pdf.

256 "Il Comitato Banda Larga", Ministero delle Comunicazioni - Ministero degli Affari Regionali c Autonomie Locali, Ministero delle Riforme e Innovazioni nella Pubblica Amministrazione, at: www.comitatobandalarga.it/news/49/10/il_comitato_banda_larga.html.

257 "Broadband and Grids Technology in the Netherlands", Ministry of Economic Affairs, 2005 at: www.hightechconnections.org/2005/broadband.pdf.

258 "Policy for the IT society: Recommendations from the members of the IT Policy Strategy Group", IT Policy Strategy Group – Sweden, 26 October 2006, at: www.sweden.gov.se/sb/d/574/a/76046;jsessionid=aRgt9J6DAf-g/.

259 "The Commission's Broadband for all" policy to foster growth and jobs in Europe: Frequently Asked Questions, European Commission Press Release, 21 March 2006 at: http://europa.eu/rapid/pressReleasesAction.do?reference=MEMO/06/132&format=HTML& aged=1&language=EN&guiLanguage=en

260 "National broadband strategy: Final report", Ministry of Transport and Communications Finland, 23 January 2007 at: www.mintc.fi/oliver/upl615-LVM11_2007.pdf.

261 "New IT Reform Strategy: Realizing Ubiquitous and Universal Network Society Where Everyone Can Enjoy the Benefits of IT", IT Strategic Headquarters – Japan , 19 January 2006, at: www.kantei.go.jp/foreign/policy/it/ITstrategy2006.pdf.

262 Informatization White Paper 2006", National Computerization Agency – Korea, September 2006, at: www.dynamicitkorea.org/koreait_policy/stat_info_9638.jsp?currentPage=1.

263 "Telecommunications Policy Review Panel Final Report 2006", Industry Canada, March 2006, at: www.telecomreview.ca/epic/site/tprp-gecrt.nsf/vwapj/00A_e.pdf/$FILE/00A_e.pdf.

264 "Stratégie du Conseil fédéral pour une société de l'information en Suisse", BACOM, January 2006, at: www.bakom.admin.ch/themen/infosociety/00695/index.html?lang=fr&download=M3wBU QCu/8ulmKDu36WenojQ1NTTjaXZnqWfVp7Yhmfhnapmmc7Zi6rZnqCkkIN1f3l7bKbXr Z2lhtTN34al3p6YrY7P1oah162apo3X1cjYh2+hoJVn6w==.pdf.

265 "Il Comitato Banda Larga", Ministero delle Comunicazioni - Ministero degli Affari Regionali e Autonomie Locali Ministero delle Riforme e Innovazioni nella Pubblica Amministrazione, at: www.comitatobandalarga.it/news/49/10/il_comitato_banda_larga.html.

266 "Information Society Strategy 2006-2010", State Planning Organisation – Turkey, July 2006 at: www.dpt.gov.tr/konj/DPT_Tanitim/pdf/Information_Society_Strategy.pdf.

267 Opinion of the Consultative Committee on Telecommunications of 31/01/2007 "How to improve broadband penetration in Belgium?", at: Dutch version: www.rct-cct.be/docs/adviezen/NL/070131%20breedband%20nl.pdf; French version: www.rct-cct.be/docs/adviezen/FR/070131%20breedband%20fr.pdf.

268 "Policy for the IT society: Recommendations from the members of the IT Policy Strategy Group", IT Policy Strategy Group – Sweden, 26 October 2006, at: http://sweden.gov.se/sb/d/574/a/76046;jsessionid=aRgt9J6DAf-g/.

269 These issues are dealt with in a forthcoming OECD paper on Convergence and Next-Generation Networks.

270 Uusien palvelujen hidasteet [Hindrances to new services] and Valokaapeli kotiin [Fibre to the home].

271 See www.culture.gouv.fr/culture/actualites/conferen/donnedieu/observatoirenum_dis.html.

Annex A

RECOMMENDATION OF THE OECD COUNCIL ON BROADBAND DEVELOPMENT

(Adopted by the Council at its 1077th Session on 12 February 2004, C(2003)259/FINAL)

THE COUNCIL,

Having regard to article 5(b) of the Convention on the Organisation for Economic Co-operation and Development of 14 December 1960;

Having regard to Rule 18(b) of the OECD Rules of Procedure;

Having regard to the document of the Committee for Information, Computer and Communications Policy entitled "Broadband Driving Growth: Policy Responses" [DSTI/ICCP(2003)13/FINAL];

RECOMMENDS that, in establishing or reviewing their policies to assist the development of broadband markets, promote efficient and innovative supply arrangements and encourage effective use of broadband services, Member countries should implement:

- Effective competition and continued liberalisation in infrastructure, network services and applications in the face of convergence across different technological platforms that supply broadband services and maintain transparent, non-discriminatory market policies.

- Policies that encourage investment in new technological infrastructure, content and applications in order to ensure wide take-up.

- Technologically neutral policy and regulation among competing and developing technologies to encourage interoperability, innovation and expand choice, taking into consideration that convergence of platforms and services requires the reassessment and consistency of regulatory frameworks.

- Recognition of the primary role of the private sector in the expansion of coverage and the use of broadband, with complementary government initiatives that take care not to distort the market.

- A culture of security to enhance trust in the use of ICT by business and consumers, effective enforcement of privacy and consumer protection, and more generally, strengthened cross-border co-operation between all stakeholders to reach these goals.

- Both supply-based approaches to encourage infrastructure, content, and service provision and demand-based approaches, such as demand aggregation in sparsely populated areas, as a virtuous cycle to promote take-up and effective use of broadband services.

- Policies that promote access on fair terms and at competitive prices to all communities, irrespective of location, in order to realise the full benefits of broadband services.

- Assessment of the market-driven availability and diffusion of broadband services in order to determine whether government initiatives are appropriate and how they should be structured.

- Regulatory frameworks that balance the interests of suppliers and users, in areas such as the protection of intellectual property rights, and digital rights management without disadvantaging innovative e-business models.

- Encouragement of research and development in the field of ICT for the development of broadband and enhancement of its economic, social and cultural effectiveness.

INVITES governments to encourage their private sector, in their broadband development activities, to take due account of this Recommendation;

INSTRUCTS the Committee for Information, Computer and Communications Policy to monitor the development of broadband in the context of this Recommendation within three years of its adoption and regularly thereafter;

INVITES the Secretary-General to make this Recommendation available to non-member economies.

Annex B
NATIONAL BROADBAND PLANS

Australia	General Information on Broadband and Internet	www.dbcde.gov.au/communications_for_business/broadband_and_in ternet
	Australian Government Information Management Office	www.agimo.gov.au
Austria	Österreichischer Forschungs- und Technologiebericht	www.bmvit.gv.at/service/publikationen/innovation/downloads/techn ologieberichte/ft_bericht06.pdf
	Sonderrichtlinie AT:net (Austrian electronic network)	www.bmvit.gv.at/telekommunikation/politik/breitband/sonderrichtlin ien/at_net.html
	IKT Masterplan 2006	www.rtr.at/web.nsf/deutsch/Telekommunikation_IKT_Masterplan/$f ile/IKT_Masterplan.pdf
		www.rtr.at/web.nsf/deutsch/Telekommunikation_IKT_Masterplan/$f ile/Masterplan_Detailanalyse.pdf
	Breitbandinitiative Österreich 2003 The Austrian Broadband Initiative	www.rtr.at/web.nsf/lookuid/4EBA719C8490EBC1C125719C00249F D8/$file/Breitbandstatusbericht.pdf
		www.bmvit.gv.at/telekommunikation/politik/breitband/index.html
Belgium	Opinion of the Consultative Committee on Telecommunications of 31/01/2007 "How to improve broadband penetration in Belgium?"	French version: www.rct-cct.be/docs/adviezen/FR/070131%20breedband%20fr.pdf".
	Policy document (October 20th, 2006) of the Minister for Economic Affairs	www.lachambre.be/FLWB/PDF/51/2706/51K2706002.pdf
	Policy document (October 13th, 2006) of the Minister for Work - part computerization	www.lachambre.be/FLWB/PDF/51/2706/51K2706001.pdf

Canada	Policy Direction to the CRTC (2006)	http://news.gc.ca/cfmx/view/en/index.jsp?articleid=219679
	Telecommunications Policy Review Panel Final Report 2006	www.telecomreview.ca/epic/site/tprp-gecrt.nsf/en/h_rx00054e.html
	Industry Canada Broadband Office and Federal Government Programs *Broadband for Rural and Northern Development (BRAND) Pilot Program *National Satellite Initiative (NSI)	http://broadband.gc.ca/pub/program/bbindex.html
	"Networking the Nation for Broadband Access" Report of the National Broadband Task Force (2001)	http://broadband.gc.ca/pub/program/NBTF/broadband.pdf http://broadband.gc.ca/pub/program/NBTF/appendix_d.html www.broadband.gc.ca/pub/program/NBTF/table_content.html
	National Broadband Task Force Report (2001)	www.broadband.gc.ca/pub/program/NBTF/table_content.html
Czech Republic	The National Broadband Access Policy 2005	www.micr.cz/files/2185/MICR_brozura_en.pdf
Denmark	IT- and Telecommunications Policy Report 2007 with description of the Danish broadband strategy and broadband related strategies	http://videnskabsministeriet.dk/site/forside/publikationer/2007/it--og-telepolitisk-redegoerelse-2007
	IT and Telecommunications Policy Report, 2006	http://videnskabsministeriet.dk/site/forside/publikationer/2006/it-and-telecommunication-policy-report-2006/Policy_report.pdf
	The Danish Approach	http://itst.dk/static/presentations/US_House_of_Representatives_210 806.ppt#265,4

EU	EU Annual Reports on i2010 Strategy	http://ec.europa.eu/information_society/eeurope/i2010/annual_report/index_en.htm
	Benchmark reports on broadband coverage in Europe	http://ec.europa.eu/information_society/eeurope/i2010/docs/benchmarking/broadband_coverage_06_2006.doc
	Broadband for all policy	http://europa.eu/rapid/pressReleasesAction.do?reference=MEMO/06/132&format=HTML&aged=0&language=EN&guiLanguage=en
	Communication - Bridging the Broadband Gap	http://eur-lex.europa.eu/LexUriServ/site/en/com/2006/com2006_0129en01.pdf
	i2010– A European Information Society for growth and employment	http://ec.europa.eu/information_society/eeurope/i2010/docs/communications/com_229_i2010_310505_fv_en.doc
Finland	National broadband strategy. Final report 2007	www.mintc.fi/oliver/upl615-LVM11_2007.pdf
	A Renewing, Human-Centric and Competitive Finland, The National Knowledge Society Strategy 2007-2015	www.tietoyhteiskuntaohjelma.fi/esittely/en_GB/introduction/
France	ARCEP's Directions about FTTH in France	www.arcep.fr/fileadmin/reprise/communiques/communiques/2007/sli-des-confpresse-ftth-281107-eng.pdf
	Plan d'actions et Forum du très haut débit	www.industrie.gouv.fr/portail/ministre/comm.php?comm_id=7137 www.internet.gouv.fr/information/information/actualites/faciliter-emergence-du-tres-haut-debit-392.html
	Plan pour une République numérique dans la société de l'information	www.premier-ministre.gouv.fr/information/fiches_52/internet_haut_debit_pour_53 101.html www.internet.gouv.fr/informations/information/plan_reso2007/ www.industrie.gouv.fr/observat/innov/lsi/dpen_planreso.pdf
	5e Comité interministériel pour la société de l'information	www.internet.gouv.fr/information/information/actualites/5e-comite-interministeriel-pour-societe-information-296.html

Country	Plan	Link
Germany	Rahmenbedingungen für eine Breitbandoffensive in Deutschland	www.diw.de/deutsch/produkte/publikationen/expublikationen/gutachten/docs/diw_rahmen_Breitbandoff200401.pdf
	German 'broadband atlas' (Broadband supply in Germany), incl. link to some studies	www.zukunft-breitband.de/Breitband/Portal/Navigation/breitbandatlas.html
	Broadband Roadshow	www.eco.de/servlet/PB/menu/1594835_11/index.html
	'German Broadband Initiative'	http://www.breitbandinitiative.de
	Initiative D21	www.initiatived21.de/
	Informationsgesellschaft 2010	www.bmwi.de/BMWi/Redaktion/PDF/Publikationen/id2010_E2_80_93informationsgesellschaft-deutschland-2010,property=pdf,bereich=bmwi,sprache=de,rwb=true.pdf
	IT Gipfel	www.bmwi.de/BMWi/Navigation/Technologie-und-Innovation/Informationsgesellschaft/it-gipfel.html
	High-Tech Strategie / IKT 2020	www.bmbf.de/pub/bmbf_hts_en_kurz.pdf www.bmbf.de/pub/bmbf_hts_lang.pdf www.bmbf.de/pub/ikt2020.pdf
Greece	The Plan for the Development of Broadband Services until 2008 Digital Strategy 2006-2013ramos	www.infosoc.gr/infosoc/en-UK/specialreports/broadband_plan/
Hungary	National Broadband Strategy	www.gkm.gov.hu/data/cms1057445/net_eng.pdf
Iceland	The policy of the Government of Iceland on the Information Society 2004-2007	http://eng.forsaetisraduneyti.is/information-society/English/nr/1327
Ireland	National Broadband Scheme	www.dcmnr.gov.ie/Press+Releases/Dempsey+Unveils+New+National+Broadband+Scheme.htm
	The Internet and Broadband Experience for Business Users	www.comreg.ie/whats_new/default.asp?ctype=5&nid=102611
	Irish Communications Market: Quarterly Key Data – March 2007	www.comreg.ie/publications/default.asp?nid=102593&ctype=5
	Next Generation Networks Forum 2007	www.comreg.ie/publications/default.asp?nid=102590&ctype=5
	Ireland Broadband Performance Review	www.forfas.ie/publications/forfas061130a/webopt/forfas061130_broadband_benchmarking_report_webopt.pdf

Italy	Interministerial Committee on Broadband development	www.comitatobandalarga.it/ www.comunicazioni.it/t/index.php?Arc=1&IdNews=234
	Infratel, company entirely owned by the Department of Economics through the National Agency for Economic Development (www.sviluppoitalia.it) to address the infrastructural digital divide in remote and rural areas.	www.infratel.gov.it
	Innovazione Italia, company entirely owned by the Department of Economics through the National Agency for Economic Development (www.sviluppoitalia.it) to address the social-economic digital divide	www.innovazioneitalia.gov.it
	Wi-max for broadband in rural areas Broadband and e-government	www.comunicazioni.it/it/index.php?IdNews=235 www.innovazionepa.it/nnovosito/pdf/linee_strategiche_egov.pdf ;
	Osservatorio Banda Larga, PPP, with the objective of measuring social and infrastructural digital divide	www.osservatoriobandalarga.it/
	Italian Regulatory Consultation on Next Generation Broadband	www.agcom.it/provv/d_208_07_CONS/d_208_07_CONS_all_B.pdf
Japan	u-Japan policy	www.soumu.go.jp/menu_02/ict/u-japan_en/index.html
	New IT Reform Strategy	www.kantei.go.jp/foreign/policy/it/ITstrategy2006.pdf
	The Next-Generation Broadband Strategy 2010	www.soumu.go.jp/joho_tsusin/broadband/index.html (Japanese only)
	White Paper - Information and Communications in Japan 2006	www.johotsusintokei.soumu.go.jp/whitepaper/eng/WP2006/2006-index.html
Korea	Broadband IT Korea Vision 2007	http://eng.mic.go.kr/eng/secureDN.tdf?seq=7&idx=2&board_id=E_0 4_03
	IT 839 strategy	http://eng.mic.go.kr/eng/index.jsp
	'u-KOREA Master Plan	www.ipc.go.kr/ipceng/policy/enews_view.jsp?num=2146

Luxembourg	The Luxembourg media and communications policy (includes regular Broadband penetration follow-up)	www.mediacom.public.lu/comm_elec/reseaux_comm/i2010_12e_rap port/index.html
	La pénétration des technologies de l'information au Luxembourg	www.entreprises.public.lu/publications/ouvrages_generaux/reperes_0 6/Reperes_2006.pdf
Netherlands	Better Performance with ICT: Update of the ICT Agenda of the Netherlands 2005 – 2006, 2005	www.micr.cz/images/dokumenty/Better_Performance_with_ICT.pdf
	Broadband and Grids Technology in the Netherlands	www.hightechconnections.org/2005/broadband.pdf
New Zealand	NZ Telecommunications Stocktake Cabinet Paper	Telecommunications Cabinet Paper (link to pdf)
	Local Loop Unbundling	www.med.govt.nz/templates/MultipageDocumentTOC____2103.asp x
	NZ's Digital Strategy	www.digitalstrategy.govt.nz/upload/Documents/MED11706_Digital %20Strategy.pdf
Norway	Report about the development of broadband in the Nordic countries	www.npt.no/portal/page/portal/PG_NPT_NO_EN/PAG_NPT_EN_H OME/PAG_NEWS?p_d_i=121&p_d_c=&p_d_v=100307
	An Information Society for All – A Report to the Norwegian Parliament	www.regjeringen.no/upload/FAD/Vedlegg/IKT-politikk/stm17_2006-2007_eng.pdf
	Homepage of the Broadband Programme Høykom	www.hoykom.no (press English)
	Theme page on ICT policy	www.regjeringen.no/en/dep/fad/Selected-topics/IT-politikk.__eNorge.html?id=1367
Poland	Draft of broadband strategy for public services for 2007-2013	www.mt.gov.pl/viewattach.php/id/1d35c10d5b152c634fd787a73342 f0d6 (Polish only)
	Polish Internet Report 2006	http://dc1.sabela.pl/raport_IAB_2006.pdf (Polish only)
	Polish fixed access market to the Internet	www.uke.gov.pl/_gAllery/54/47/5447/Dostep_szerokopasmowy_do_ sieci_Internet.pdf (Polish only)
	Informatization Strategy of Poland until 2013	www.mswia.gov.pl/download.php?s=1&id=1652 (Polish only)

Portugal	Technological Plan: A national strategy for growth and competitiveness based on knowledge, technology and innovation.	www.planotecnologico.pt
	Connecting Portugal: An action plan, for the Information Society, included in Technological Plan.	www.ligarportugal.pt
Slovak Republic	The National Broadband Access Policy	www.ufe.cz/research/projects/breath/events/Event7_documents/Murin-Scehovic.pdf
Spain	Ingenio 2010	http://www.ingenio2010.es
	Avanza Plan	http://www.planavanza.es
	National Program of broadband infrastructures deployment in rural and isolated areas	http://www.bandaancha.es www.bandaancha.es/EstrategiaBandaAncha/ProgramaExtensionBandaAnchaZonasRuralesAisladas/EnglishInformation/
Sweden	Policy for the IT society - Recommendations from the members of the IT Policy Strategy Group	www.sweden.gov.se/sb/d/574/a/76046
	From an IT policy for Society to a Policy for the Information Society, 2005	www.regeringen.se/sb/d/108/a/47411
	Broadband for growth, innovation and competitiveness	www.sweden.gov.se/sb/d/574/a/76048
	Proposed Broadband Strategy for Sweden	www.pts.se/Dokument/dokument.asp?ItemId=6541
Switzerland	Universal Service Obligation 2008 – 2017	www.bakom.ch/dokumentation/medieninformationen/00471/index.html?lang=en&msg-id=7308
	La stratégie du Conseil fédéral pour une société de l'information en Suisse	www.bakom.ch/themen/infosociety/00695/index.html?lang=fr
	Le Rapport 2007 du Comité interdépartemental chargé de mettre en oeuvre la stratégie du Conseil fédéral pour une société de l'information	www.bakom.ch/themen/infosociety/index.html?lang=fr

Turkey	Broadband Strategy	www.abgs.gov.tr/tarama/tarama_files/10/SC10DET_10A-BroadbandStrategy.pdf
	Information Society Strategy and its Action Plan	www.bilgitoplumu.gov.tr/eng/docs/Information_Society_Strategy.pdf www.bilgitoplumu.gov.tr/eng/docs/Action_Plan.pdf
United Kingdom	The Government's broadband policy	www.dti.gov.uk/files/file37744.pdf
United States	FCC broadband resources	www.fcc.gov/cgb/broadband.html
	FCC semi-annual report, "High-Speed Services for Internet Access"	www.fcc.gov/wcb/iatd/comp.html
	White House document, "New Generation of American Innovation (Promoting Innovation and Economic Security through Broadband Technology)," April 2004.	www.whitehouse.gov/infocus/technology/economic_policy200404/innovation.pdf
	"Broadband Rights-of-Way Memorandum," April 2004.	www.whitehouse.gov/news/releases/2004/04/20040426-2.html
	President's November 2004, Executive Memorandum concerning "Improving Spectrum Management for the 21st Century."	www.ntia.doc.gov/osmhome/spectrumreform/Nov2004PresidentialMemo.pdf
	National Telecommunications and Information Administration, Networked Nation: Broadband in America, 2007, released January 2008	www.ntia.doc.gov/reports/2008/NetworkedNationBroadbandinAmerica2007.pdf

Source: Submisssions of OECD country delegations.

OECD PUBLICATIONS, 2, rue André-Pascal, 75775 PARIS CEDEX 16
PRINTED IN FRANCE
(93 2008 02 1 P) ISBN-978-92-64-04668-9 – No. 56221 2008

Printed in the United Kingdom
by Lightning Source UK Ltd.
134610UK00001B/360/P

9 789264 046689